CONSTRAINT THEORY

MULTIDIMENSIONAL MATHEMATICAL MODEL MANAGEMENT

More information about this series at http://www.springer.com/series/6104

CONSTRAINT THEORY

MULTIDIMENSIONAL MATHEMATICAL MODEL MANAGEMENT

Second Edition

George J. Friedman • Phan Phan

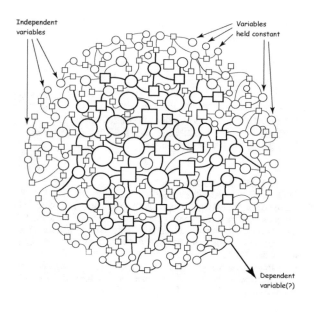

Independent variables

Variables held constant

Dependent variable(?)

Springer

George J. Friedman
Viterbi School of Engineering
University of Southern California
Los Angeles, CA, USA

Phan Phan
Viterbi School of Engineering
University of Southern California
Los Angeles, CA, USA

ISSN 1574-0463
IFSR International Series on Systems Science and Engineering
ISBN 978-3-319-85483-0   ISBN 978-3-319-54792-3   (eBook)
DOI 10.1007/978-3-319-54792-3

Printed on acid-free paper

This Springer imprint is published by Springer Nature
The registered company is Springer International Publishing AG
The registered company address is: Gewerbestrasse 11, 6330 Cham, Switzerland

Fronticepiece

24" Doily, designed and constructed by Regina Oberlander (Dr. Friedman's mother) circa 1915 in Chinedeev, near Muncacevo, Austro-Hungarian Empire.

A single circuit cluster with over 1,000 independent simple circuits (thus there are 1000 more edges than vertices)

This book is dedicated to our wives, who have helped us find the time and priority for our mathematical dreams.

Preface

At first glance, this might appear to be a book on mathematics, but it is really intended for the practical engineer who wishes to gain greater control of the multidimensional mathematical models which are increasingly an important part of his environment. Another feature of the book is that it attempts to balance left- and right-brain perceptions; the authors have noticed that many graph theory books are disturbingly light on actual topological pictures of their material.

Constraint Theory was originally defined by George Friedman in his PhD dissertation at UCLA in 1967 and subsequent papers written over the following decade. There was a dearth of constraint theory publication after the 1970's as Dr. Friedman was working on several classified aerospace programs wherein publication of any kind was most difficult. The first edition of this book was published in 2005. Constraint Theory was further extended by Phan Phan in his PhD dissertation at USC in 2011, leading to this second edition.

Acknowledgments

Over the past several years, Constraint Theory has been a substantial part of a special graduate course in the Systems Architecture and Engineering program at the University of Southern California, where the authors are adjunct faculty members. The feedback from this group of bright graduate students was invaluable. Special thanks are given to these graduate students who performed research studies – summarized in Appendix A – on Constraint Theory: Leila Habibibabadi, Kathya Zamora-Diaz and Elliott Morgan [1]. Extra special thanks are given to Mr. Gary L. Friedman who provided the extremely valuable service of intellectual reflector, editor and digital format manager for this book.

Introduction

Many thousands of papers have been written about the accelerating pace of increased complexity and interactivity in virtually every walk of life in the developed world. Domains which previously could have been studied and managed separately – such as energy, the environment and economics – must now be dealt with as intimately intertwined disciplines. With its multitude of additional capabilities, complex systems also provide a treacherous array of fragile failure modes, and both the development and operation of new systems are an increasing challenge to the systems engineer. Advanced technology is the primary driving force behind the increasing complexity and the enthusiastic pushing of immature technologies is behind most of the early failures in the development phases.

Perhaps the most significant advanced technology employed in new complex systems is the computer science family with its ancillary disciplines of communications and software. Fortunately, computer science also represents a major opportunity to control the design and operation of complex systems because of its ability to perform multi-dimensional modeling to any level of detail desired. Math models have been used in support of every phase of systems engineering, including requirements management, preliminary design, interface and integration, validation and test, risk management, manufacturing, reliability and maintainability, training, configuration management and developing virtual universes to measure customer preferences prior to the implementation of the design. Properly used, the enormous power of modern computers can even furnish researchers with a synthetic world where theories can be tried and tested on validated models, thus avoiding far more expensive tests in the real world. A wide variety of questions – or "tradeoffs' – can be asked of the models and, at least in theory, the analyst has a free choice as to which computations he wishes to observe and which variables he desires to be independent. Philosophically, it can even be argued that the math model employed in this fashion provides the technologist a virtual extension of the scientific method itself.

Those who have actually wrestled with large-scale models will complain that the above description is far too rosy. Submodels which are developed by separate organizations are normally very difficult to integrate into an overall working model; they often must be dealt with as "islands of automation." The greatest of care must be taken to make sure that the definition of each variable is clear and agreeable to every member of the team. In general, it is difficult to distinguish between a model and the computer program, and if a computational request is made which reverses dependent and independent variables, then the model must be reprogrammed. To say the least, much diligent effort must be undertaken to obtain the many advantages promised by mathematical modeling.

However, even after the diligence, there exists a much deeper problem that often diminishes the utility of math modeling; it is associated with the traditional "well posed" problem in mathematics. We need to know whether the model is internally consistent and whether computational requests made on it are allowable. The alarming facts are that models constructed by diverse teams – and this is normally the case for very large models – have internal inconsistencies and that most of the possible computational requests which can be made on even consistent models are not allowable. This problem is the domain addressed by Constraint Theory and is the subject of this book.

Chapter 1 provides an example of low dimension, showing how problems of consistency and computational allowability can arise in even simple situations. The reader is introduced to the two main characters of the book – an experienced manager and an analyst – whose dialogue will hopefully illuminate the book's many concepts. The bipartite graph is introduced, as are a few simple rules. However, the analyst argues that, in order to expand the tools to models of very high dimension, and in order to trust the reliability of these tools, the theory must be based on a more rigorous foundation. "Only the simplest 5% of graph theory and set theory are required", he claims.

Chapter 2 begins to establish the rigorous foundation by defining four "views" of a mathematical model: 1) set theoretic, 2) submodel family, 3) bipartite graph, and 4) constraint matrix. The first two views are full models; the last two views are metamodels. Then, rigorous definitions of consistency and computational allowability are made in the context of these views.

Chapter 3 discusses the similarities between language and mathematics and provides some general consistency and computability results with respect to any class of relation. In order to provide a basis for the next three chapters, three classes of exhaustive and mutually exclusive relations are defined: discrete, continuum, and interval.

Due to the amount of new materials resultant from Dr. Phan's research as part of his doctoral dissertation, previous Chapter 4 in the first edition has been split into two new Chapters 4 and 5 in this second edition. Chapters 4 and 5 represent the core of constraint theory at its present stage.

As before, the new Chapter 4 addresses the constraint theoretic properties of *regular* relations, the most important type within the continuum class, and the most often employed in the development of multidimensional math models. The simple rules presented in Chapter 1 are rigorously proved employing the foundations of Chapters 2 and 3. The topological properties of the bipartite graph are analyzed to provide key conclusions of the model's consistency and computational properties.

A specific type of subgraph within the bipartite graph, called the *Basic Nodal Square (BNS)* is identified as the "kernel of intrinsic constraint" and is accused of being the culprit in model inconsistency and unallowable computability. Trivially easy computations on the bipartite graph – such as circuit rank and constraint potential – are shown to have enormous utility in locating the BNSs which hide in tangled circuit clusters.

Additionally, the newly updated Section 4.6 extends, in more prescriptive details with graphical illustrations, the step by step algorithm for locating BNSs within a model graph.

The new Chapter 5 discusses the general issue of constraint propagation through a connected model graph of regular relations. A detailed procedure for determining model consistency and computational allowability in such a model is introduced.

In particular, Section 5.4 introduces new mathematical definitions and theorems to enable the detection of overlapping BNSs. Section 5.5 outlines techniques to relieve over-constraint among them. Section 5.6 describes a precise procedure to expand their resultant constraint domains. Section 5.7 prescribes, in details with graphical illustrations, a step-by-step algorithm to process computation requests made on a model. And Section 5.8 provides a constraint theory toolkit to employ the rules and theorems in an orderly manner and which can find BNSs trillions of times faster than brute force approaches.

Chapter 6 addresses the constraint properties of *discrete* and *interval* functions such as those from Boolean algebra, logic and inequalities. These classes of relations are less important in support of modern math modeling, but strangely, it was the first that the author studied in his development of Constraint Theory. It was easier for him to imagine multidimensional sets of points than multidimensional sets of continuous functions. Interval relations require the greatest interaction between models and metamodels, and the concept of constraint potential is less useful than for regular relations.

Chapter 7 provides a compact structure of constraint theory. All postulates, definitions and theorems are listed and their logical interrelationships are displayed in the form of quasi-bipartite graphs.

Chapter 8 presents detailed examples of the application of constraint theory to the areas of operations analysis, kinematics of free-fall weapon delivery systems and the dynamics of deflecting asteroids with mass drivers.

Chapter 9 summarizes the book and provides the manager and analyst a final opportunity to dialogue and discuss their common background.

Problems for the interested student are presented at the end of most chapters, so this book could be employed as a text for a graduate course – or senior level undergraduate course – in Systems Engineering or mathematical modeling.

Of course, a complete list of references is provided, as well as an index.

Several appendices treat detailed material to a depth that would slow down the natural rhythm of the exposition if they were included in the chapters themselves. Appendix A is noteworthy in that it summarizes the research projects on "computational request disappointments." On models approximately the size of Chapter 1's "simple example" – eight variables – the percentage of allowable computational requests based on the total number of possible computational requests is only on the order of 10%. It is presently "Friedman's conjecture" that as the dimensionality, K, of the model increases, the number of allowable computational requests also increases, perhaps as fast as the square of the model's dimension or K^2. However, the number of possible computational requests increases far faster: 2^K. Thus, for a 100-dimension model, only 10^{-26} of all possible computational requests will be allowable! Models of thousands of dimensions have been built and are planned; so the ratio of allowable to possible computational requests is enormously worse that even this incredibly low number. The technologist who wishes to gain maximum benefit from asking his model to perform any computation his imagination conjures up will certainly be disappointed! A tool such as constraint theory which will lead him to the 10,000 computational requests ($K=100$) or 1,000,000 requests ($K=1,000$) which are allowable should be valuable.

Appendix B provides a very brief overview of graph theory with the objective of justifying why the bipartite graph was chosen as the primary metamodel for constraint theory.

Appendix C describes the rigorous logic of the difference between "if and only if" and "if" types of theorems. Most of constraint theory's theorems are of the latter category – a source of confusion to many students.

The newly updated Appendix D establishes fundamental algebraic structures which are essential to implement constraint theory. These include definitions and properties of general vector spaces, and binary set operations.

A Warmup Problem in Complexity

This book makes substantial use of a mathematical structure from graph theory called a bipartite graph. In the past, bipartite graphs have been employed to solve "pairing" problems associated with various social situations such as picnics or dinner parties.

Out of respect for this tradition, let us consider a set of five men – named Jack, Jake, Jude, Juan, and Jobe – and a set of five women – named Jane, Joan, June, Jean, and Jenn. Let us define a *relationship pattern* as a complete description of all heterosexual relationships between the five men and five women. For example:

- In the *communal* pattern, every man has a relationship with every woman. There is one such pattern.
- In the *celibacy* pattern, none of the men have a relationship with any of the women. Again, there is one such pattern.
- In the *male harem* patterns, one of the men has a relationship with each of the women, but all the other men are devoid of relationships, except perhaps to be eunuchs. There are five such patterns. Similarly, there are five possible *female harem* patterns.
- In the *monogamy* patterns, each man has a relationship with exactly one woman and vice versa. There are 5!=120 such distinct patterns.

And so on. There are many more patterns. The question is: *What is the total number of possible heterosexual relationship patterns between five men and five women?*

The answer – discussed in Chapter 4 and Appendix A – may surprise you: it's over 30 million (!). It certainly surprised the author and changed an important objective of his research agenda. Moreover it represents the hidden depths possible in apparently simple problems of low dimension as well as a challenge to one's belief in intuition or rational mathematics.

About the Authors

George Friedman is a Professor of Practice in the Astronautical Engineering Department of the Viterbi School of Engineering of the University of Southern California. He has developed and taught graduate courses in systems engineering with emphasis on the management of complexity and decision science. This book is the product of one of these courses.

He has had over 45 years of experience in industry, retiring from the Northrop Corporation as their Corporate Vice President of Engineering and Technology. He worked on a wide variety of aerospace programs and served as a consultant to all branches of the Department of Defense, NASA, the National Science Foundation, and Department of Energy as well as to the NATO industrial advisory group.

He was a founder of the International Council on Systems Engineering (INCOSE), served as its third president, was elected a fellow and is on the editorial board of INCOSE's journal, *Systems Engineering.*

He has also been a member of the Institute of Electrical and Electronic Engineers (IEEE) since its formation from the IRE and AIEE, was elected a fellow and was the vice president for publications of the *IEEE Transactions on Aerospace and Electronics Systems.* He received the Baker Prize for the best paper published by all societies of the IEEE in 1970 – the subject of the paper was Constraint Theory.

He was a former director of research at the Space Studies Institute at Princeton, and had supported several new technologies associated with the long range development of space.

He received the Bachelor's degree in engineering at the University of California at Berkeley and the Masters and Doctorate at UCLA. The topic of his PhD dissertation was constraint theory [2, 3].

Phan Phan is a Lecturer in the Astronautical Engineering Department of the Viterbi School of Engineering of the University of Southern California (USC). He has assisted and taught graduate courses in systems engineering, systems management, lean operations and economic analysis.

He has had over 36 years of technical and managerial experience in government, military and various industries, including oil & gas exploration, commercial and military aircraft, unmanned sensors, and major weapon systems.

His industry assignments included Mobil Research & Development, General Dynamics, Lockheed Aeronautical Systems, McDonnell Douglas and Boeing Integrated Defense Systems. As a registered Professional Engineer in California, he currently works as a reliability analyst with the Naval Surface Warfare Center – Corona Division.

He has also served in the U.S. Navy Reserve as an Engineering Duty Officer, attaining the rank of Captain, and currently assigned to Naval Sea Systems Command. His previous Navy assignments included Office of Naval Research/Naval Research Laboratory, Program Executive Office Integrated Warfare Systems, Naval Space & Warfare Systems Command, Mobile Mine Assembly Group, Naval Weapons Station Seal Beach and Naval Shipyard Long Beach.

He received his B.S. in Engineering from the University of Alabama in Huntsville, Master of Engineering from the University of Texas at Arlington, MBA from California State University – Fullerton, M.S. in Systems Architecture & Engineering from USC, Master of Engineering Acoustics from the Naval Postgraduate School, and Ph.D. in Industrial & Systems Engineering from USC.

The topic of his doctoral dissertation was "Expanding Constraint Theory to Determine Well-Posedness of Large Mathematical Models" [22], the main contribution to the second edition of this book.

Contents

Chapter 1 **MOTIVATIONS**

What is Constraint Theory and why is it important?

1.1 TRENDS AND PROBLEMS IN SYSTEM TECHNOLOGIES

Gone forever are the simple days! Virtually every identifiable trend is driving humanity's enterprises into more intimate interaction and conflict. Increased population, accelerated exploitation of resources, and expanded transportation have brought the previously decoupled worlds of economics, energy and the environment into direct conflict. With the greater efficiency of travel and communication, the emergence of global marketplaces and the revolution in military strategies, the international world is incredibly more interactive, multidimensional and complex than even a decade ago. Locally, we observe ever tighter coupling between emerging problems in crime, poverty, education, health and drug misuse. All these issues have been aggravated by an explosion of new technology and – especially in the United States – a compulsion to force these new technologies into early and often simultaneous application. The most vigorous of these advancing technologies – digital computation – brings with it an unexpected complexity challenge: software and the management of highly complex and multidimensional mathematical models.

Fortunately, this most rapidly advancing technology of computer science not only adds to the complexity of *designed* systems, it also contributes enormously to *designing* these systems themselves. A host of new "computer assisted" software packages are published each year, running the gamut from Computer Assisted Design (CAD), Computer Assisted Engineering (CAE), Computer Assisted Systems Engineering (CASE), Computer Assisted Manufacturing (CAM), Computer Assisted Instruction (CAI), and eventually

© Springer International Publishing AG 2017
G.J. Friedman, P. Phan, *Constraint Theory*, IFSR International Series on Systems
Science and Engineering, DOI 10.1007/978-3-319-54792-3_1

to Computer Assisted Enterprise Management (CAEM). This family of tools permits the design engineers to control a virtually unlimited number of variables, to predict behaviors and performance of systems still in their early conceptual stages, to optimize with respect to detailed criteria, to effect interdisciplinary integration and to perform design changes with unprecedented speed and accuracy. It is not an exaggeration to claim that without this array of computer-based tools, many systems that exist today would have been impossible to design and implement.

However, as most systems engineers who attempt to gain benefit from these tools are well aware, computer-based design is a mixed blessing. A common complaint is that the various programs which support facets of the total problem are "islands of automation" – they are difficult to integrate into a total system problem solving capability. Another problem is that the tools are virtually useless in sorting out the variety of languages and technical shades of meaning, especially on highly interdisciplinary systems. Yet another challenge for the engineering and program managers is the vigorous and frequent upgrading of every hardware and software package causing unprecedented costs of initial installation and training to the overall design process, not to mention the inevitable bugs in the early versions. Many companies have even established new organizations whose members are expert in computer-assisted programs, and not expert in the technical design itself.

Observers who watched the agonizing entry of computers over the last several decades into many diverse worlds such as financial management, stock market trading, airline ticketing, air traffic control, and education should be optimistic that eventually computer-based design will also become an efficient tool which will become so easy to use that investing in it will be clearly justified. But in order for this dream to occur, there are more problems to solve – deeper problems than getting the definitions sorted out and software packages to work together.

Even when the willing but cognitively challenged computational giants of computer based tools are completely manageable, several fundamental problems will still exist, mostly on the cognition and mathematical levels.

Nothing is said in any of the present set of textbooks on Systems Engineering about the regrettable "subdimensionality" of the human intellect. Despite the fact that a thorough description of a modern complex system requires the understanding and integration of hundreds to thousands of variables, cognitive scientists have known for decades that the human mind is limited in its perceptive powers to a mere half-dozen dimensions. Regardless of all our other miraculous gifts such as language, art, music, imagination, judgment, and conscience, our dimensional perceptive power is tiny compared to the challenges of designing complex systems, and sadly, it appears that this is a "wired in" shortcoming of our nervous system and thus we cannot hope to be trained to attain a higher perceptive capability. This "dimensionality gap" is severely aggravated by the habit many self-styled "decisive managers" have in further suppressing their limited perception by searching for simplifications such as "the bottom line," "the long pole in the tent" or "getting the right angle" in attempting to make complex descriptions more comprehensible to them. Typically, when the dimensionality of the model overwhelms that of the decision maker, and he sees results which appear to be anti-intuitive, he will tend to distrust these results as the product of software bugs or other errors. Thus, major opportunities to learn from the enhanced power of modeling are lost because the operation of the computer becomes more and more opaque to the decision maker as the dimensionality increases.

As an aside, we humans also have problems with numbers: we cannot perceive 29 in the same way we perceive 5. The raw arithmetic perception of the average person is "the magic number seven, plus or minus two," according to the cognitive scientist George Miller. However, after a lifetime of dependable experience with arithmetic algorithms, we have the illusion that we can understand and manage entities such as 29, or even 29,000,000,029.

The other fundamental problem was previously referred to as the "well-posed" problem in mathematics. That is, when a mathematical model is established, is it internally consistent? When computational requests are desired based on this model, are they allowable? If the answer to either question is "no," then we have a situation which is not "well posed" and we can expect nonsensical results or jammed up attempts to program. This problem is made worse by the fact that in most digital computer programs, models are built with a unidirectional computational flow that was anticipated by the programmers, but is not necessarily responsive to the needs of the decision makers. It was a source of great irritation to this author to be told many times over his career that a computational request was "impossible" because the model was programmed with another computational flow in mind. However, when the reprogramming *was* done in an attempt to be more responsive, more fundamental problems frequently arose.

An example of these problems will be useful at this point. The example given in the next section was chosen to be as simple as possible, but still indicating aspects of the well posed problem that can arise even without our entering a dimension so high that our perceptions are boggled.

1.2   AN EXAMPLE OF LOW DIMENSION

A decision-making manager was authorized to initiate the preliminary design of a new system development by his board of directors. In the true spirit of systems engineering, he realized the importance of making the best decisions as early in the system development process as possible. Accordingly, he gathered a team of the best specialists available, along with a systems analyst to help organize the math model that he hoped would guide him to strategic systems tradeoffs and decisions.

The chief systems engineer stressed that, in order for an "optimum design" to exist, it was necessary to define a total systems optimization criterion, T:

$$T = PE/C \tag{1}$$

where:
   P was the political index of acceptability by the board of directors,
   E was the system effectiveness, and
   C was the life cycle cost of the system.

The operational chief, expressing a weariness with the overly aggressive use of new and unproved technology on most of his previous systems, wanted to stress that most of the total system cost should be applied to operations and support, not new systems development. Thus, he contributed this limitation:

$$D = k_1 C, \text{ where } k_1 = 0.3 \tag{2}$$

where D, the development cost, was to be limited to 30% of the total cost.

The operations and support specialist, attempting to predict the level of cost after production and delivery were complete, provided:

$$S = X + 0.5D \tag{3}$$

where:
   S is the total support cost
   X is the cost of ops and support if there were no new technology
   D is the development cost for the system, including new technology.

The systems costing and estimating specialist contributed the obvious:

$$C = D + S \qquad\qquad (4)$$

taking care that all ambiguities between development, operations and support costs were clearly defined and resolved.

The reliability, maintainability and availability specialist provided:

$$S = K_2 E / (1-A) \qquad\qquad (5)$$

where:

K_2 is a constant,
A is the probability that the system is ready when called upon.

Finally, the operations analyst provided this definition of effectiveness:

$$E = MA(D/D_{max})^{1/2} \qquad\qquad (6)$$

where:

M is the mission success probability, given the equipment is available
D is the amount spent on development
D_{max} is the budget requested by the developers

All these inputs from the specialists were reviewed for reasonableness by the systems analyst and integrated into the "model" shown in Table 1-1.

Table 1-1. The Mathematical Model

1) $T = PE/C$
2) $D = k_1 C$ where $k_1 = 0.3$
3) $S = x + 0.5D$
4) $C = D + S$
5) $S = k_2 E / (1-A)$
6) $E = MA(D/Dmax)^{1/2}$

where:
T = Top-level systems criterion
P = Political index of acceptability
C = Life cycle cost of a system
D = Development costs of system
S = Support and operations cost
E = Effectiveness of system
M = Mission success probability (working)
A = Availability of system

"There appear to be no internal inconsistencies," reported the analyst to the manager. "Indeed, this model is enormously simpler than any I have ever dealt with for years."

The manager, who claimed many years of systems engineering experience, observed, "I see the model is imbedded in an eight-dimensional space and is constrained by six equations. Therefore, there should be two "degrees of freedom." Since I'm most concerned with the total system optimization criterion, please compute plots of $T = f_7(S,P)$ for me."

"Sorry, said the analyst, that is not an allowable computation on this model. Although the *total* model seems to have two degrees of freedom, that freedom does not exist uniformly throughout all parts of the model. In particular, the submodel composed of relations 2, 3 and 4 is concerned only with the variables C, D and S. Therefore, in the three-dimensional subspace of CDS, we have three equations and three unknowns; thus there are <u>no</u> *degrees of freedom*, and these variables are constrained to a point or a set of points. Since it is such a constrained variable, S obviously cannot act as an independent variable for the computational request, $T = f_7(S,P)$."

The manager did not like the word, "obviously". "There must be something wrong with the model," he asserted. The specialists got huffy.

The analyst assured the manager, "There are no inconsistencies or internal contradictions in this model. Once we've agreed to accept some inaccuracy due to simplification, all the equations are 'correct' and perfectly valid mathematically. Each of the relations referring to CDS space was contributed by a separate specialist. Because of this interaction between three disciplines, C, D and S are determined and can no longer be considered as variables. Any computational request which includes C, D or S as an independent variable must be considered unallowable. Your request was not mathematically well-posed."

"All right," conceded the manager, "then let me see $T = f_8(M,A)$."

"Sorry again," said the analyst, "that request is also not allowable. Consider the relations 5 and 6 which are concerned with variables S, A, E, D and M. As we've just discussed, S and D are held to constant values because of the *internal* constraint applied from another part of the model. By applying M and A as independent variables, we are applying *external* constraint to the SAEDM space. Thus, we have only one variable, E, which is neither internally or externally constrained, and which must conform to the two equations, 5 and 6. Having two equations with only one unknown is a clear case of local *over*constraint. This computational request is also not well posed."

The manager sighed, "Then what computations *are* allowable of the form, $T = f(p,q)$?"

The analyst replied, "only these three: $T = f_9(E,P)$, $T = f_{10}(M,P)$ and $T = f_{11}(A,P)$. All other computations of the form $T = f(p,q)$ either overconstrain or underconstrain some part of the model."

These computations were plotted and given to the manager, along with the constant values for C, D and S. After studying these results, the manager said, "OK, next I'd like to see some tradeoff curves. Please show me the tradeoff between M and P, everything else being equal."

"By 'everything else being equal' do you mean: 'hold all other variables at constant values?'" asked the analyst.

"Yes, I suppose so," responded the manager, fearing the worst.

"In that case the desired tradeoff is not allowable," said the analyst. "Once we agree that C can no longer be considered as a variable, E is the only variable that connects the submodel containing M with the submodel containing P. If E is held to a constant value as you want, then the two submodels are essentially disconnected and the M vs. P tradeoff cannot be computed."

"What tradeoffs *would* be allowable?"

"If we hold T only at a constant value, then M vs. P, A vs. P and E vs. P are allowable computations."

After all the allowable computations were performed and examined, the manager asked, "How were you able to come to your conclusions on the various computational allowabilities so rapidly? Do you have a method that provides you special insight?"

The analyst showed the manager Figure 1-1. "Fundamentally, I attempt to get a right-brain view of the topology of the model. Look at relation 5 for example. In Figure 1-1a, I represent the relation by a square and show its three relevant variables S,E,A, – represented by circles – connected to it. Note that there are no arrowheads on these connections, since we don't know in advance what the computational path will be. In Figure 1-1b, the arrowheads indicate that S can be computed if A and E are known inputs. Similarly, Figures 1-1c and 1-1d show the computational flow directions for the computations of A and E respectively."

"Now, let's expand our perspective to include relation 6 and its relevant variables, A,E,M and D. Looking at Figure 1-2, we see that variables A and E are common to both relations 5 and 6; they do not have to be repeated. Note that the topology has developed a little circuit."

"Continuing in this manner, we can include the entire model, shown in Figure 1-3. This structure is called a bipartite graph which provides a right-brained view of the *topological structure* of the model of Table 1-1. It's really a *metamodel* since it does not contain the actual equations of the original model – just the structural information necessary to determine internal consistency and computational allowability. As in all bipartite graphs, there are two distinct types of junctions: squares represent the relations, and circles represent the variables. The arcs, or "graph edges" connect each relation to the variables that are relevant to it. The "degree," d, of each junction is defined as the number of edges which intersect it."

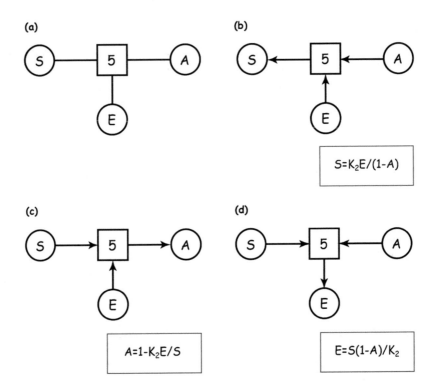

Figure 1-1. A right-brain view of relation 5, introducing the concept of representing relations by squares and variables by circles, as well as demonstrating that computation can flow through the relations in many directions.

"This bipartite graph can be considered as a network for information flow. The squares are essentially multidirectional function generators – or algorithmic processors – such that any output can be generated if all the other edges provide input. The circles are essentially scalar measurements of the value of the variable that they represent."

"The above use of the bipartite graph for the representation of a math model can be easily extended to the representation of a computational request. In the general format of a computational request, one specifies a dependent variable (the output) and a set of independent variables and variables held constant (the input). These input variables essentially have constraint applied to them – in addition to the constraint applied to them by their relevant relations – and thus additional squares are appended to the bipartite graph.

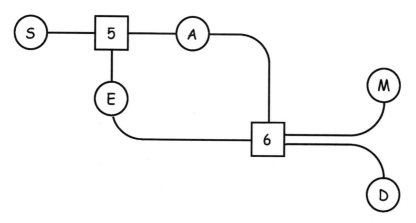

Figure 1-2. The perspective is expanded to include both relations 5 and 6, which share variables A and E.

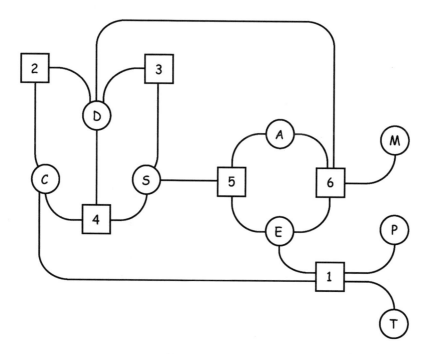

Figure 1-3. The Bipartite Graph: A Metamodel displaying consistency and computability.

For example, assume that it is desired to have variables M and P be independent variables and variable A be held constant. As is shown in Figure 1-4, the squares identified with "I" are appended to M and P, while the square identified with "C" is appended to variable A. In summary, the squares representing relations of the model imply *intrinsic* constraint, while the squares representing inputs to a computational request apply *extrinsic* constraint. To emphasize this difference, the intrinsic constraint squares have a single border while the extrinsic constraint squares have a double border."

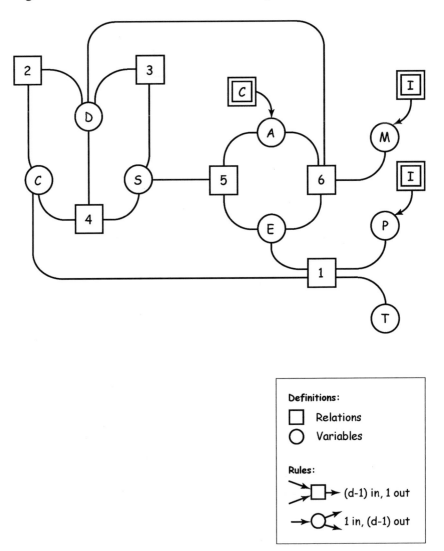

Figure 1-4. Symbology and rules of propagation for bipartite model graph.

"Before a computational request is made, the edges of a bipartite graph model have no directionality. Once the request is made, the input variables now apply constraint to their neighbors and the edges take on a directionality which is determined by the request. Essentially, the computation or constraint "flows" across the graph."

"For treelike graphs, the rules to map the sequential propagation of information – or computation, or *constraint* – are simple in the extreme: for d edges intersecting a square, information will propagate if there are (d-1) inputs and one output; for d edges intersecting a circle, information will propagate if there is one input and (d-1) outputs." (See Figures 1-1 and 1-4.) Note that a square with but a single arc intersecting it will automatically transmit constraint to the circle it is connected with, since in this case, d=1 and (d-1)=0, thereby requiring no inputs to generate its output.

"For graphs which contain circuits, mapping propagation is a little more complex. Once we have gone as far as we can with the *sequential* rules above, we may require the rule of *simultaneous* propagation in the vicinity of a circuit: if a connected subgraph exists which contains an equal number of unpropagated squares and circles, then all its variables may be computed as if they were within a set of simultaneous equations."

"Now I can show you how easy it was to determine the computability of your computational requests. Looking at Figure 1-5, we can easily see that relations 2, 3 and 4 form a submodel with an equal number of squares and circles. This denotes three simultaneous equations covering three unknown variables and we should expect to be able to solve for the three variables, converting them from unknown variables to fixed parameters. Thus, your request, $T = f_7(S,P)$ is unallowable since it assumes S is variable rather than a fixed parameter. For the same reason, C and D cannot be independent variables either. This type of constraint imposed on the model is called *intrinsic* because it existed even before you made any computational request."

"Now look at Figure 1-6, which shows constraint propagating from left to right along variables C, D and S for the above explained reason. When you requested the computation $T = f_8(M,A)$ you established the two independent variables as a source of *extrinsic* constraint which propagates into the model and hopefully gives us a computation of T. Let's see what happens when we employ the sequential propagation rules for constraint flow. Since M and A are extrinsic sources of constraint and D is an intrinsic source, we can satisfy the (d-1) inputs and one output rule for equation 6, thus producing a computation for variable E. With A as an extrinsic source and S as an intrinsic source, we can satisfy the (d-1) inputs and one output rule for equation 5, thus producing *another* computation for variable E. This is a case of local overconstraint, making the requested computation unallowable."

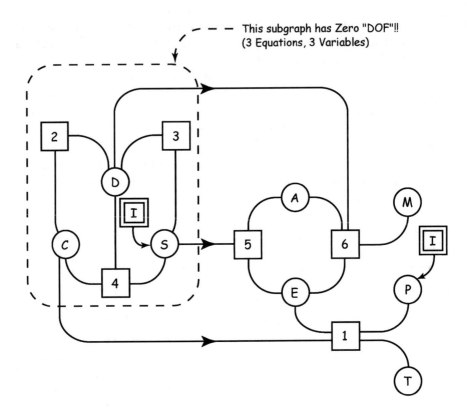

Figure 1-5. T = f₇(S,P) is <u>not</u> allowable.
S cannot be an independent variable.

"Don't get discouraged, allowable computations exist also. Figure 1-7 displays the computational paths to compute T = f₉(E,P). Note how the extrinsic constraint flows from E and P combine with the intrinsic flow from C to satisfy the (d-1) inputs and one output rule to equation 1, resulting in the computation of T. Also note that the entire model was not necessary for this computation, as equations 5 and 6 were irrelevant to it."

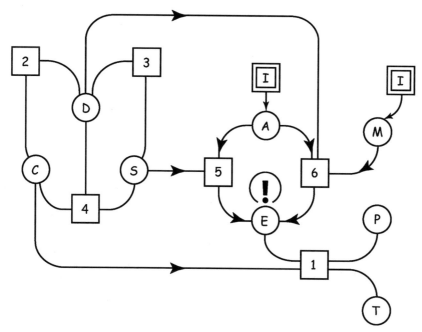

Figure 1-6. T = f₈(M,A) is not allowable
because of overconstraint on E.

"Figure 1-8 shows a case where both sequential and simultaneous computational flow is used to satisfy the request, T = f_{10}(M,P). As originally constructed, the submodel comprised of the equations 5 and 6, together with their relevant variables S,A,E,D and M had two equations and four variables – 'two degrees of freedom' as you put it. Now, with the application of intrinsic constraint from S and D and extrinsic constraint from M, the extra degrees of freedom collapse to zero and we are left with two equations and two variables for this submodel. We can expect to solve these two simultaneous equations in two unknowns to obtain both A and E. Now applying the intrinsic constraint flow from C, the simultaneous constraint flow from E and the extrinsic constraint flow from P to equation 1, we can compute T. Thus, this request is allowable. In this case, the entire model was involved with the computation."

"Figure 1-9 shows the computational paths for the allowable request T = f_{11}(A,P). The intrinsic input from S with the extrinsic input from A permit equation 5 to compute E which, using the 1 in, (d-1) out rule, propagates to both equations 6 and 1. This input from E plus the intrinsic input from C and the extrinsic input from P permits 1 to compute T as requested. By the way, this same bipartite graph shows that M = f_{12}(A) is also allowable."

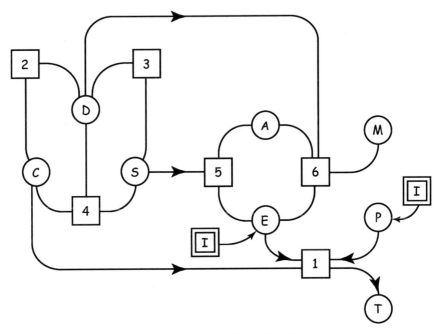

Figure 1-7. T = f_9(E,P) is allowable.
The constraint flow rules work OK.

"Now the term, 'tradeoffs, with everything else held equal' is fraught with ambiguity and most often used with insufficient rigor. Figure 1-10 displays how holding E at some constant value effectively decouples M and P into different, non-communicating subgraphs, rendering this tradeoff request unallowable.

If any combination of C, D and S were to be held constant, two types of problem emerge. First of all, the value to which they were held constant might not agree with the values computed from the 2, 3 and 4 simultaneous equations. Even if they did agree, then propagating constraint across the graph would yield an *under*constraint at equation 1 – there would be an insufficient number of inputs to provide the desired computation of P. The same underconstraint situation occurs if the variable held constant is A. In fact, the *only* variable that can be held constant in order to provide the M vs P tradeoff is T."

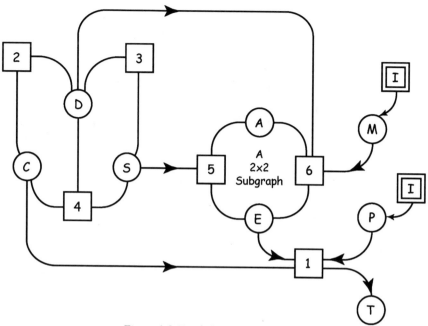

Figure 1-8. T = f_{10}(M,P) is allowable.
A new BNS is formed; then the flow is OK.

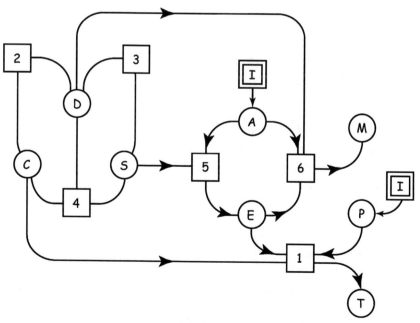

Figure 1-9. T = f_{11}(A,P) is allowable.
M = f_{12}(A) is also allowable.

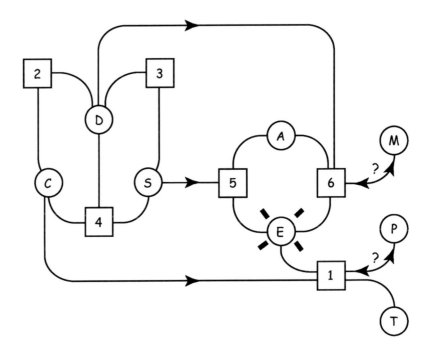

Figure 1-10. M = f$_{13}$(P) "Tradeoff" is not allowable
holding E constant decouples M and P is allowable.

1.3   THE MANAGER AND ANALYST CONTINUE
THEIR DIALOGUE

The manager absorbed all these inputs soberly and after reviewing the
results of his requested computations as well as others which were allowable
on the original model, he complained, "I certainly can't argue with your
mathematical rigor. But I'm still disappointed that I didn't get more insight
out of this model for my preliminary design phase – I expected more,
somehow. It seems that whether a computation is allowable or not is like the
flip of a coin."

"It's worse than a coin flip; much worse. You have just observed a very
common and generally unappreciated feature of most math models," the
analyst responded. "Indeed, the vast majority of all possible computational
requests on almost all models are unallowable. Some of the author's
graduate students performed an exhaustive analysis of the likelihood of
computational allowability on 6- and 8-dimensional connected models of a
variety of topologies. The results are presented in detail in Appendix A, but
the general allowability likelihoods were surprisingly low. Of the 150

computational requests made on the 6-dimensional model, less than 15% were allowable. Of the 1078 computational requests made on the 8-dimensional model, less than 6% were allowable. As the dimensionality of the model increases to dozens or hundreds of variables to address modern complex systems, then we can expect the allowability likelihood to diminish even further. Needless to say, this is far worse than a coin flip."

"But don't lose hope, you can still 'negotiate' a more useful model with your team of specialists – after all, they have an even more limited systems view than you about the structure of the model and know even less about your intended use of it."

"For example, let's examine the three relations, 2, 3 and 4, that caused the inadvertent source of intrinsic constraint. Equation 4 is merely a definition between three types of cost – this certainly seems OK. Equation 3 is the result of experience of how support costs increase with new technology developments – this is OK based on the experience of many past systems and assumes that there will be no investment in the development phase to reduce supportability costs. But now look at equation 2; it is not a representation of a definition or an experienced relationship, it is a *policy* statement by a person who is attempting to limit development costs so he can spend more on operations and support. If, instead of demanding that $K_1 = 0.3$, he permitted K_1 to merely be another variable in the model, the model dimensionality would increase to nine, and more importantly, the intrinsic source of constraint due to equations 2, 3 and 4 would be relieved (see Figure 1-11). By this reasonable negotiation, we increase the candidates for independent variables to include C, D, S and also K_1. In fact, the operational chief who furnished equation 2 in the first place can now run studies to determine what value of K_1 will maximize the systems level criterion, T, rather than arbitrarily fixing it at 0.3."

"Very interesting," admitted the manager, "I can see a constructive integration between left- and right-brain views. It would be extremely difficult to initiate negotiations of this type without the visibility provided by your bipartite graphs.

I'm surprised I haven't seen this methodology before. But most real world problems are vastly more complex than this example, are they not? Wouldn't analysts be driven crazy if they tried to work with snake charts that were many square meters in area?"

"You're absolutely correct," agreed the analyst. "Meaningful models quickly get large and even rigorous graphs become like a bundle of snakes, as you put it, and not really amenable to the analysis by inspection shown in the example. Figure 1-12 displays the flow graph for a specific computational request on a model that's about π times larger than the example. The author actually was able to do consistency and allowability analysis on this size model without using computer aids. This was possible because the rules, and later, the theorems were developed using comprehensible models of low dimension, and then extending them rigorously to higher dimensionality. For example, the model depicted in Figure 1-13 – again about another π times larger than Figure 1-12 – would undoubtedly boggle the mind of even the most focused analyst unless computer aids were available. In order to communicate with the computer, a new construct – called the 'constraint matrix', which has all the information inherent in the bipartite graph – will be employed."

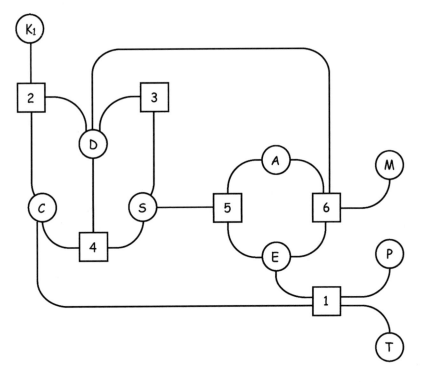

Figure 1-11. Changing K_1 from a constant to a variable permits C, D, S, and K1 to act as independent variables in computations.

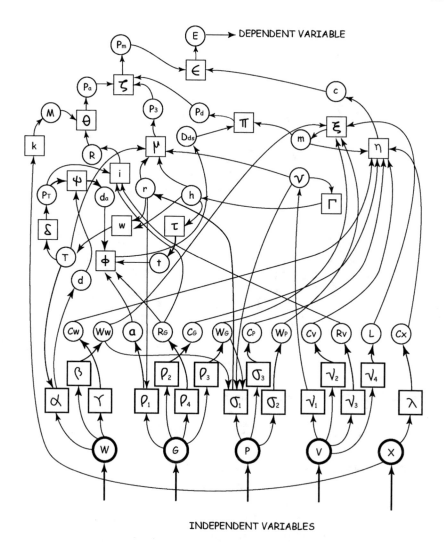

Figure 1-12. A sequential computational flow in a model about π times larger than the example.

"Is there a name to this process? How difficult would it be to become proficient in this technique? What would the mathematical prerequisites be?" asked the manager.

"The name of the process is 'Constraint Theory'. It is based on the author's PhD dissertation [2] and subsequent published papers [3]. The only mathematical prerequisites are the simplest 5% of set theory and graph theory. Just read this book; it is written for practical engineers, not professors or journal editors."

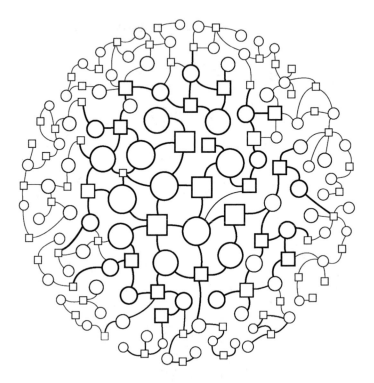

Figure 1-13. A model about another π times larger certainly invites computer assistance.

1.4   PRELIMINARY CONCLUSIONS

Today's dynamic world of systems development is virtually in an explosion of complexity and multidimensionality. Computer science represents our best hope of controlling the complex multidimensionality; however a major barrier to its trustworthy use is the "well posed problem."

Constraint Theory addresses the two fundamental issues of the well-posed problem:

(a) Is a mathematical model internally consistent?
(b) Are computational requests made of the model allowable?

Even with simple 6- or 8-dimensional models, the vast majority of computational requests are not allowable from a well-posed standpoint.

Constraint Theory also provides decision-making managers greater visibility into the assumptions and structure of the contributing relations underlying models and a basis for negotiating alterations to the model in order to attain greater benefit through desired computations.

Constraint Theory's only mathematical prerequisites are elementary set theory and graph theory.

Compared to the present practice of digital simulation which tightly integrates the mathematical model with rigid directions of algorithmic flow, Constraint Theory distinguishes the mathematical model from computational flow and permits multidirectional flow at the request of the analysts.

The bipartite graph and its companion, the constraint matrix, provide insightful topological metamodels to both the model and its computations. They provide a viewpoint to establish a rigorous mathematical extension of the methodology to any number of dimensions.

1.5 A LITTLE WINDOW INTO FUTURE CHAPTERS[1]

For those systems engineers and managers who wish to gain
MULTIDIMENSIONAL INSIGHT
almost comparable to their perceived understanding of
NUMBER AND ARITHMETIC
and learn the activities of
NORBERT WIENER BEFORE CYBERNETICS
and
CLAUDE SHANNON BEFORE INFORMATION THEORY
and ponder the deep nature of
RELATIONS
with their many diverse characterizations including
BIPARTITE GRAPHS AND OTHER RIGHT BRAIN METAMODELS
which assist in solving the venerable
WELL-POSED PROBLEM
by detecting and correcting the dual villains
OVERCONSTRAINT AND UNDERCONSTRAINT
and ferreting out the chief culprit of inadvertent constraint
THE BASIC NODAL SQUARE
from its hiding place deep within
HOPELESSLY TANGLED CIRCUIT CLUSTERS
by employing easy graph theoretic measurements of
CIRCUIT RANK
and
CONSTRAINT POTENTIAL
which locate these kernels of constraint within medium-sized models
TRILLIONS OF TIMES
more rapidly than would a random search or the use of Hall's 1914
theorem,

THEN DO AS THE ANALYST SUGGESTS:
READ THE BOOK!

---

[1] Inspired by Edgar Palmer's [4] delightful title page for his book, *Graphical Evolution,* John Wiley and Sons, New York, 1985

1.6 PROBLEMS FOR THE CURIOUS READER

1. Provide three examples from your knowledge or experience of small models being developed and then integrated into total system models.

2. Provide four examples of the utility of computers in the design, production or operation of complex systems, where humans would find it difficult or impossible to handle.

3. Provide five examples of problems associated with computers applied to complex systems.

4. Regarding the mathematical model depicted in Figure 1-3, which of the following computational requests are allowable and which are not allowable?

   For the allowable requests, draw the directed bipartite graph which depicts the computational flow. For the unallowable requests, discuss the reason(s) for the unallowability.

   Computational Requests:
   M=f(A), M=F(E), E=f(A,M), P=f(T,E), P=f(T,A)

5. For those computational requests which you deemed to be allowable, switch the dependent variable with one of the independent variables and check again for allowability. Does this suggest a possible generalization?

Chapter 2 THE FOUR-FOLD WAY

How to Perceive Complex Mathematical Models and Well-Posed Problems

2.1   PROLOGUE: THE MANAGER AND ANALYST DISCUSS THE ORIGINS OF MULTIDIMENSIONAL MODELS AND WELL-POSEDNESS

"Since complexity has grown so enormously in modern times," the manager commented, "I presume that the motivations to develop techniques to manage it are relatively recent."

"On the contrary," replied the analyst, "many of the concepts and examples of problem recognition are quite old – ancient even."

Consider the old Indian story of the blind men trying to "understand" an elephant. Depending on what is touched – the leg, ear, tail, trunk, or tusk – the unknown object takes on the attributes of a tree, a leaf, a rope, a snake or a spear. Thus, touching an aspect of a complex object is far removed from understanding the total integrated concept of "elephant."

A more recent story – but still almost 2000 years old – comes from the Talmud [5]. According to a commentary on the book of Genesis, on the day that the Lord created Man, He took truth and hurled it to the ground, smashing it into thousands of jagged pieces. From then on, truth was dispersed, splintered into fragments like a jigsaw puzzle. While a person might find a piece, it held little meaning until he joined with others who had painstakingly gained different pieces of the puzzle. Only then, slowly and

G.J. Friedman, P. Phan, *Constraint Theory*, IFSR International Series on Systems Science and Engineering, DOI 10.1007/978-3-319-54792-3_2

deliberately, could they try to fit their pieces of Truth together – to make some sense of things.

Mankind's yearning to understand the world over the eons has been aided by the development of mathematical models. Groups of researchers, sometimes spanning centuries contribute their little fragments of data or understanding and eventually a general theory emerges. In many cases, the consequences of the new theory are unexpected by the original contributors, but such is the trust given to mathematics, the unexpected, nonintuitive results are accepted given they are mathematically sound. Examples:

- In the 16th century, Tycho Brahe organized and extended the astronomical observations of Copernicus and others into the world's finest set of data on stellar and planetary objects. Johann Kepler took this data and formulated his famous three laws of planetary motion. Despite his disappointment that planetary orbits were elliptical – rather than the circles the Greeks maintained were necessary for "celestial perfection" – he convinced himself and the scientific world that the ellipse was the correct mathematical form for all the orbits in Tycho's data base.
- Decades later, Isaac Newton, with his greater mathematical understanding, was able to generalize Kepler's laws into his law of universal gravitation – a gigantic intellectual feat which unified the laws of the heavens and earth.
- Centuries later, Albert Einstein provided a refinement of Newton's theory of universal gravitation with his general theory of relativity. Alexander Friedmann solved Einstein's equations and concluded that the universe began in a monstrous big bang. This was so against Einstein's instincts that he added a cosmological constant to his equations of relativity to remove the possibility of an expanding universe or the big bang. However, the rationality of mathematics, as well as new data by Hubble and others established Friedmann correct and Einstein has referred to the cosmological constant as his greatest blunder.

So in Man's quest to understand, mathematical modeling has taken an increasingly central role in building theories, and indeed in the scientific method itself. The jagged shards of data, incomplete observations and subdimensional theories are pieced together rationally – often resulting in unexpected conclusions and a deeper view of the world. With the advent of modern computer technology, this central importance promises to increase far more.

"You certainly won't get arguments from most practitioners of science and technology about the importance of computers," remarked the manager, attempting to be agreeable. "What you have said would be obvious to most observers."

"What is not obvious is that there are many barriers to the future efficient use of computers in the modeling of complex system," rebutted the analyst.

"I knew you'd say that," said the manager, remembering the example of Chapter 1. "What are these barriers?"

"First of all," began the analyst, "with all the increased capability and flexibility that the digital computer offers over the analog, there comes a subtle but pervasive disadvantage: the model and the computational requests placed on it are inextricably intertwined. In almost all cases, the model is programmed to execute a specific computational flow, and when asked to alter the computation or switch input and output variables on the same model, the programmers tend to tell the managers, "can't be done" or "too much trouble, or "can't you make do with all that I've given you?"

"Amen," agreed the manager, "I've been told that many a time. The programmers love to overwhelm you with data to show off their powerful computation. Their love of being responsive to your deep needs to understand what the model is teaching us is unfortunately much less."

Second, until early this century, the general concept of a relation has been quite fuzzy and philosophical. Then in 1913, Norbert Wiener [6], before he became the father of cybernetics, suggested that the definition of a relation be imbedded within set theory – one of the foundations of all mathematics. This served to add needed clarity and rigor to the concept of "relation."

Third, there was a general expectation that once a model was developed, there were no limitations on what computations could be asked of it. Which questions are "well posed" and which are not? In 1942, Claude Shannon [7], before he became the father of information theory, studied these issues on the recently developed mechanical differential analyzer – the most powerful computer of its time, analog or digital. He discovered that some of the variables desired to be dependant, or output variables – based on the rotation of a shaft assigned to that variable – were "free running", providing no useful results. In other cases the entire network of rotating shafts, gear trains and integrators would just "lock up" – again providing no useful results. These instances of "free running" and "lockup" are directly related to the concepts of under constraint and over constraint, which we will discuss later.

Fourth, as was mentioned in Chapter 1, there is a vast dimensionality gap between the cognitive capability of man and machine. Our challenge is to make the best partnerships between these cognitive entities. As George Gamow [8] related in his charming book, "One, Two, Three, Infinity," it was possible to survive with very limited numerical perceptions during our primitive beginnings, but the advent of mathematics, starting with arithmetic, enormously enriched our lives and ability to understand and control the world.

"I can see the issues are not as new as I thought," admitted the manager.

"Speaking of all the fathers, you mentioned Gamow – wasn't he also a student of Alexander Friedmann, the father of the big bang theory as well as the father of George Friedman, this book's author?"

"Almost correct! You continue to amaze me, I should develop more respect for you," beamed the analyst. "Yes, George Friedman's father was Alexander Friedmann, but he was Friedmann the tailor, not Friedmann the cosmologist. But let's proceed to some substance."

"OK," challenged the manager. "I'm ready to enter the mathematical world you tell me that is necessary to bring order to this confusion and ambiguity. Let's see if the work will prove to be a worthwhile expenditure of intellectual energy."

"Fair enough," agreed the analyst. "In the remainder of this chapter, I want to introduce you to the very simplest foundations of set theory and graph theory, which will define for us with rigor and clarity the formerly vague concepts of relation, well-posed, consistent, allowable computation, overconstraint and underconstraint. I believe it will be worth your effort."

We will begin our exploration of the foundations of constraint theory by presenting four interrelated views of the mathematical model: set theoretic, families of submodels, bipartite graph, and constraint matrix. The first and second are complete and contain all the model's detail. The third and fourth are metamodels and contain only those abstractions which illuminate the model's structure as it relates to consistency and computability.

2.2   THE FIRST VIEW: SET THEORETIC

Definition 1: A *set* is a collection of elements. A *subset* is a portion of this collection. The number of elements may be finite, such as the planets of the solar system, or infinite, such as the points on a line. A set with no elements at all is the null set. (Figure 2-1)

Definition 2: A *variable* is an abstraction of one of the model's characteristics which the analyst considers essential. Associated with each variable is an *allowable set of values*. (Figure 2-2)

The set of variables which define the model can have enormous flexibility. The variables can be continuous and quantitative, such as force, length, or temperature; they can be discrete, such as the variables in Boolean Algebra, or the solutions of Diophantine equations; they can be qualitative, such as hot, rich, salty or sick; or combinations of these.

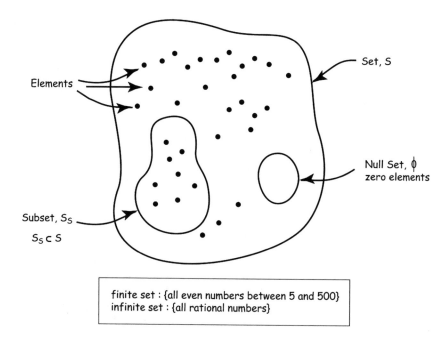

Figure 2-1. It all begins with the simple concept of sets, subsets, and the null set.

VARIABLE TYPE	ALLOWABLE VALUES
Money	Any real value
Length, Mass...	Positive Real Numbers
Gear Ratio	Rational Numbers
Probability	0 < real number < 1
Binary Boolean	0 or 1
Ternary Boolean	0, 1 or 2
Permutations	Positive integers
Taste	Sweet, Sour, Salty, Bitter
H_2O State	Ice, Water, Steam
Binary Relations	Larger than, Older than...

Figure 2-2. The sets of variables and their allowable values have enormous flexibility.

Definition 3: The *model hyperspace* is that multidimensional coordinate system formed by all the variables as axes, each of which is orthogonal to all the others. (Figure 2-3) This is simply a generalization of Descartes' unification of geometry and algebra. We will frequently refer to these axes as Cartesian coordinates.

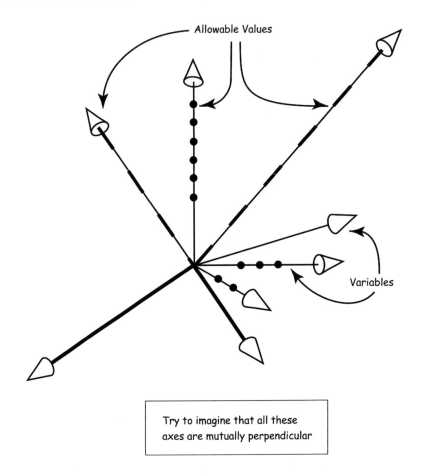

Figure 2-3. The Model Hyperspace formed by the orthogonal axes of the variables is a useful abstraction, although in general it is impossible to be perceived.

Definition 4: The *product set* of a set of variables is the set containing all possible combinations of the allowable values of all the variables. In the case where all the variables are continuous over an infinite range, the product set is merely every point within the hyperspace defined by the set of variables. (Figure 2-4)

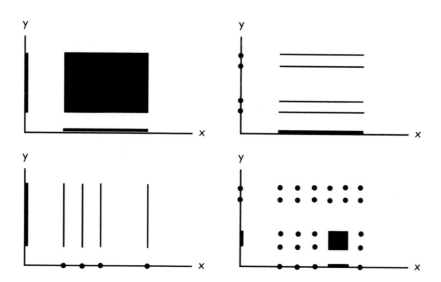

Figure 2-4. The Product Set contains all possible combinations of the allowable sets of the variables' values.

Definition 5: As suggested by Wiener and amplified by Bourbaki [9] and Ashby [10], a *relation* between a set of variables is defined as that subset within the product set of the variables which satisfies that relation. (Figure 2-5) This relation can be between any number of variables and is not restricted to the binary relations of "relation theory."

The relations can also have enormous flexibility. They can be linear or nonlinear, differential equation, partial differential equations, integral-differential equations, logical equations, binary, ternary, etc., deterministic or probabilistic, inequality relations, or any combination of these. In many cases the relations can be represented by data or "truth tables."

Definition 6: Since a relation reduces the size of the original product set to a smaller, relation set, the relation can be said to *constrain* or apply a *constraint* to the original product set. (Figure 2-6)

Now that we have embedded the concept of mathematical models within set theory, we will need these four set theoretic operations for further developments: (see definitions 7 and 8; Figure 2-7)

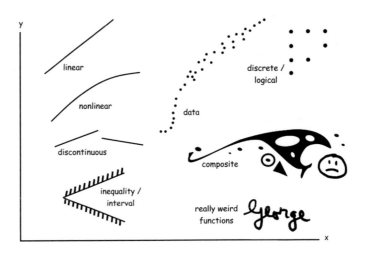

Figure 2-5. Wiener suggested that a "relation" between variables can be defined as the subset within the product set of these variables which satisfies it. This not only provides rigor, but permits a tremendous variety of relation types.

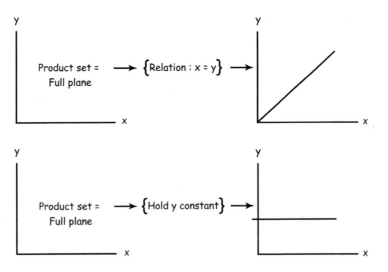

Figure 2-6. Relations, as well as variables held constant, constrain the product set into a much smaller subset.

Union:

> {A} ∪ {B} = all elements in A or
> B or both

Intersection:

> {A} ∩ {B} = all elements that are
> in both A and B

Projection:

> $Pr_y A$ = the shadow of A onto y
> $Pr_x A$ = the shadow of A onto x

Extension:

> ExA = extends each point in A
> to all points in the new direction

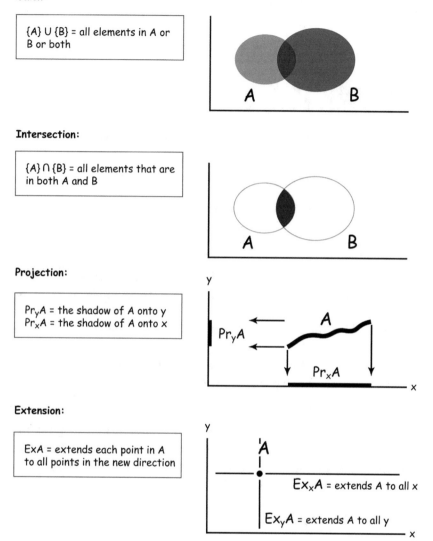

Figure 2-7. Only four operations from set theory are employed for Constraint Theory.

<u>*Definitions 7*</u>: The *union* of sets A and B is the set of all points which are either in set A or in set B or both. Symbolically:

$$x \subset A \cup B \text{ if: } x \subset A \text{ <u>or</u> } x \subset B$$

The *intersection* of sets A and B is the set of all points which are in both A and B. Symbolically:

$$x \subset A \cap B \text{ if: } x \subset A \text{ \underline{and} } x \subset B$$

Definitions 8: The *projection* of set A onto dimension x is the set of points within set A with all coordinates except x suppressed. For example, if set A is the point (2,4) in the xy plane, then Pr_x (2,4) = 2 on the x axis only. Projection is a dimension reducing operation. Symbolically:

$$\text{If } A = (2,4) \text{ in xy-space} \qquad \text{(point)}$$

$$Pr_xA = (x{=}2) \text{ in x-space} \quad \text{(point)}$$

$$Pr_yA = (y{=}4) \text{ in y-space} \quad \text{(point)}$$

The *extension* of set A into dimension y is the set of all points within set A plus all possible values of the dimension y. For example, if set A is the point (2,4) in the xy plane, then $Ex_y(2,4)$ is the line $x{=}2$ where y varies over all its possible values. Extension is a dimension increasing operation. Symbolically:

$$Ex_yA = (x{=}2) \text{ in xy-space} \quad \text{(line)}$$

$$Ex_zA = (x{=}2) \cap (y{=}4) \text{ in xyz-space} \quad \text{(line)}$$

Definition 9: y is a *relevant* variable with respect to relation ϕ in *xyz* space means that there exist lines in *xyz* space parallel to the y axis that are neither entirely within nor entirely outside of the relation set. Thus y has an effect on ϕ, or equivalently, the relation ϕ constrains y. Symbolically:

$$\text{If: } Ex_y(Pr_{xz}A_\phi) \neq A_\phi, \text{ then y is relevant to } \phi$$

Similarly, y is irrelevant with respect to relation ϕ if:

$$Ex_y(Pr_{xz}A_\phi) = A_\phi \quad \text{(Figure 2-8)}$$

2.3 THE SECOND VIEW: FAMILY OF SUBMODELS

The set theoretic definition of relation was chosen to provide the firmest and broadest mathematical foundation for the work to follow. Unfortunately, it cannot also be claimed that this viewpoint is a practical way to *describe* the relation. There are some occasions, such as tabulated or plotted functions, when it is necessary to list every point within the relation subset exhaustively. In these cases, the relation subset is merely the union of all the listed points within the hyperspace of the model. However, in the vast majority of mathematical models, far more efficient means are used to define the usually infinite number of points comprising the relation subset.

These efficient means almost invariably involve the concept of describing the total model as the intersection or union (or both) of a set of submodels or algorithms. (Figure 2-9) This is necessary for at least two reasons: First, and more obvious, a practical way of specifying infinite sets is required. Second, and deeper, model builders cannot conceive of the entire model with their limited perceptual dimensionality and thus attempt to construct higher dimensional models by aggregating in some fashion a series of lower dimensional submodels. The rules of aggregation employ the union, intersection, projection and extension operators defined previously.

Frequently, a function is specified in a piecewise fashion; for example:

$$x = 0 \text{ when } t<0$$
$$x = t^2 \text{ when } t>0$$

In cases like this, the meaning is that the contribution of these two sets to the total model is the *union* of the sets.

More frequently, a collection of "simultaneous equations" attempt to define the model; for example:

$$x + y + z = 13$$
$$x - y = 8$$

In cases like this, the meaning is that the contribution of these two sets to the total model is the *intersection* of the sets.

In general, the dimensionality of the total model is far greater than any of the contributing submodels. Thus, the contributing submodels specify only a subset of the total model and, in order for them to be able to intersect in the total dimensional space, they must be *extended* into all the unspecified directions. For example, let the total model space be *xyz* and let the relation subset for $f_1(x,y)=0$ be A_1 and the relation subset for $f_2(x,z)=0$ be A_2. Thus, before these two relations intersect, A_1 must be extended in the missing z direction, and A_2 must be extended in the missing y direction. Defining A_Σ as the total model relation, then:

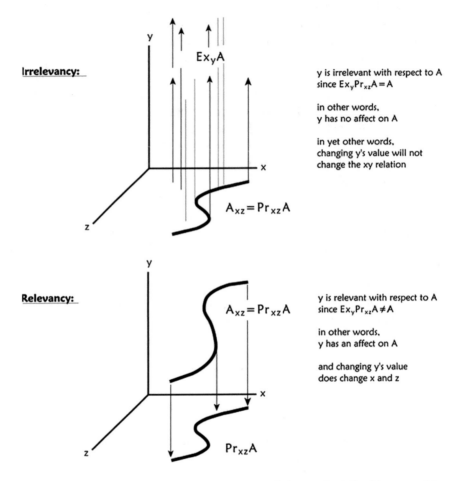

Irrelevancy:

y is irrelevant with respect to A
since $Ex_y Pr_{xz} A = A$

in other words,
y has no affect on A

in yet other words,
changing y's value will not
change the xy relation

Relevancy:

y is relevant with respect to A
since $Ex_y Pr_{xz} A \neq A$

in other words,
y has an affect on A

and changing y's value
does change x and z

Figure 2-8. Relevancy of a variable with respect to a relation can be defined in terms of the projection and extension operations.

$$A_\Sigma = Ex_z(A_1) \cap Ex_y(A_2)$$

Now, once the model is constructed in the above fashion, an analyst wishes to have a subdimensional "view" – or computational request – of this multidimensional relation. In order for him to view the relation – as A_V – with respect to the *xy* plane, he must ask for a projection of A_Σ onto the *xy* plane. (Figure 2-10). Symbolically:

$$A_V = Pr_{xy}(A_\Sigma)$$

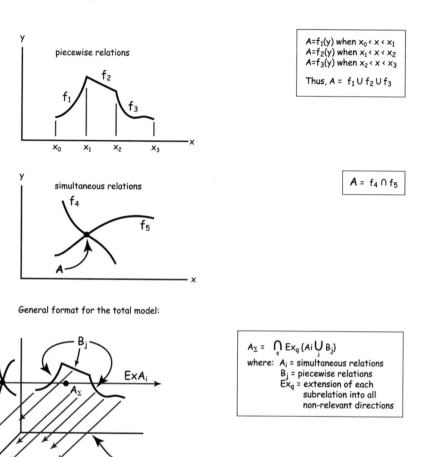

Figure 2-9. Total model relations are generated by families of submodels which are combined by the union, intersection, extension and projection operations.

If the analyst wishes to impose additional restrictions on his view, or computation, prior to the projection, he may hold any number of variables at a constant value. In these cases, the relations corresponding to these variables held constant intersect the total model relation prior to the application of the projection operation.

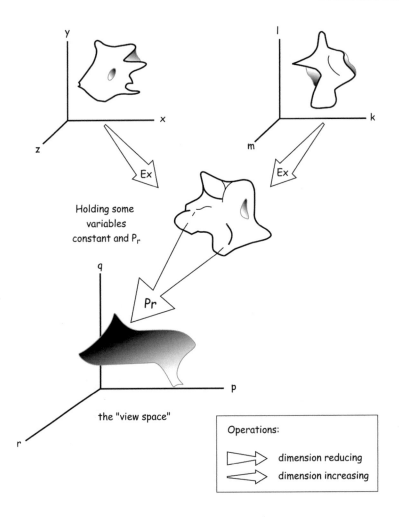

Figure 2-10. After each generating relation is developed, it is *extended* into the full hyperdimensional space of the total model, forming the relation A_Σ, which in turn is *projected* onto the subspace of the computational request, where it can be "viewed" by managers, analysts and others interested in learning from the model.

2.4 THE THIRD VIEW: THE BIPARTITE GRAPH

Although there exist strong implications of topological structure in mathematical models and their computations, neither of the two views described above provides topological insight. In order to provide this additional insight – as well as allow a right-brain perspective to aid the dominantly left-brain views already presented – graph theory will be applied.

Definitions 10: A *graph* is a topological network of points, called *junctions*, or *vertexes,* and lines connecting some of them, called *arcs,* or *edges.* A *bipartite graph* is a special graph having two disjoint sets of vertexes, *{K}* and *{N},* such that any edge is only allowed to connect a vertex in *{K}* to a vertex in *{N}.*

Definitions 11: A *model graph,* is a bipartite graph with one set of vertexes, called *nodes,* corresponding to the model's relations and the other set of vertexes, called *knots,* corresponding to the model's variables. A knot will be connected by an edge to a node only if the corresponding variable is relevant to the corresponding relation. As an additional visual aid, nodes will be shown as squares and knots will be shown as circles. (Figure 2-11)

A model graph can be thought of merely as the circuit diagram of a computer hookup of the mathematical model: the nodes are function generators and the knots are wired connections that permit the values of the variables to pass from one function generator to another. Thus, when the edges indicate no direction, the bipartite graph represents a model. When the edges indicate specific directions, then the bipartite graph represents a computation on that model, tracking the flow of computation or constraint across the topological structure.

2.5   THE FOURTH VIEW: THE CONSTRAINT MATRIX

The fourth and final viewpoint of the mathematical model is introduced primarily to provide a format amenable to computer processing. As will be seen later, however, it also furnishes yet another mathematical perspective from which the proof of certain theorems can most easily be made.

Definitions 12: A *constraint matrix* is a rectangular array of elements that presents exactly the information inherent in a bipartite model graph, but is a form that can be easily stored and operated upon by a computer. The columns correspond to variables and the rows correspond to relations. An element in the ith column and the σth row will be filled if the variable i is relevant to the relation σ, and empty if it is not. (Figure 2.12) Compactly stated, the rows, columns and elements of the constraint matrix are homomorphic to the nodes, knots and edges of the bipartite graph. In order to indicate the direction of computational flow, the elements of the constraint matrix can take on the values: +1 or -1.

To further emphasize the essential similarity between the bipartite graph and the constraint matrix, Figure 2-13 shows a logical evolutionary transition between the two representations. As was stated earlier, both the bipartite graph and constraint matrix are "metamodels" and do not contain the full model information inherent in the set theoretic and family of submodels versions. Rather, they emphasize the structure and topology important to model consistency and computational allowability.

2.6   MODEL CONSISTENCY AND COMPUTATIONAL ALLOWABILITY

We are now prepared to present rigorous definitions in the area of the "well posed problem." Saying a problem is well posed means that the mathematical model is consistent and the computation is allowable.

Definition 13: A mathematical model is *consistent* means that its multidimensional relation set contains at least one point. Symbolically,

$$A_\Sigma \neq \textbf{ the null set}$$

Definition 14: A computational request made on a model is *allowable* means that the projection of A_Σ onto the view space of the computation contains at least one point and in addition, each variable involved in the computation must be relevant to this projection in the sense of definition 9.

Thus, if the projection onto the desired subspace that the analyst wants to view has been nulled out to no points at all, then the computation, the application of variables held constant or even the total model relation has been *overconstrained*. On the other hand, if the projection has variables that are not relevant, these variables take on all their possible values, and are therefore *underconstrained*. (Figure 2-14)

2.7   THE MANAGER AND ANALYST CONTINUE THEIR DIALOGUE

"You started off simply enough," commented the manager. "What can be easier than the definition of sets and their operations? Also, your extension of Cartesian coordinates into hyperspace can be grasped by extrapolating from what we know of one, two and three dimensional spaces.

As a teenager, I was inspired by Abbot's *Flatland* [11] and Burger's *Sphereland* [12]. These extensions are interesting philosophically, but do they really represent the real world and should they be the basis for applied mathematics? I've heard professors argue that Descartes himself only was thinking of our familiar three dimensional space, not even the four dimensions for relativity theory, the eleven dimensions for string theory, and certainly not the hundreds of dimensions we need for a modern mathematical model."

Graph:

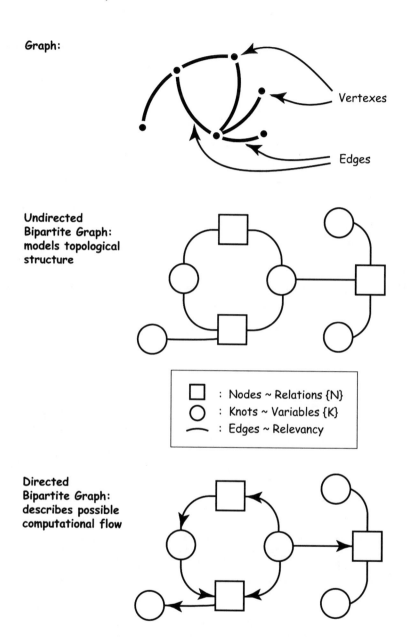

Undirected Bipartite Graph:
models topological structure

☐ : Nodes ~ Relations {N}
○ : Knots ~ Variables {K}
— : Edges ~ Relevancy

Directed Bipartite Graph:
describes possible computational flow

Figure 2-11. The bipartite graph is a metamodel of the full model which illuminates the model's structural and computational properties.

Bipartite Graph {node, knot, edge} :: Constraint Matrix {row, column, element}

The Constraint Matrix displays structure

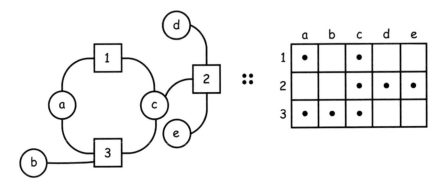

> ● denotes that the variable in the column is relevant to the relation in the row

The Constraint Matrix displays computation (or constraint flow)

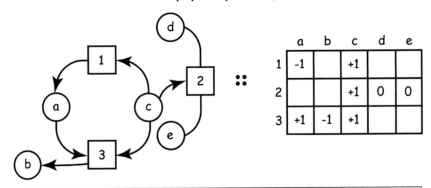

> +1 denotes constraint flow from knot to node
> -1 denotes constraint flow from node to knot
> 0   denotes no constraint flow from, although relevancy exists

Figure 2-12. The Constraint Matrix is the companion to the bipartite graph and displays exactly the same information. It is also a metamodel which contains only that information relating to the model's structural and computational properties.

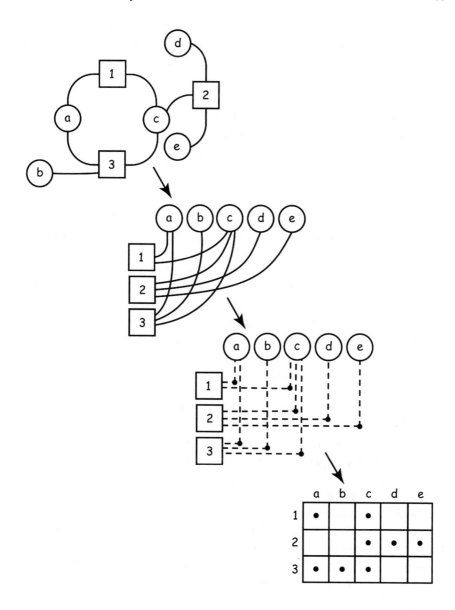

Figure 2-13. Evolutionary transition from the bipartite graph to the constraint matrix.

"What do we really know about 'reality,'" asked the analyst, launching into a minor tirade. "Irrational numbers were once thought to be irrational, and still bear the label. Negative numbers were once thought to be imaginary, but now play an essential role in every walk of science and finance. Zero was originally thought to be completely unworthy of serious consideration, but we would be crippled if we stuck to the awkward Roman numerals. Imaginary numbers were originally considered as only interesting abstractions, but today form the basis of several very practical integral transforms and are essential to almost every walk of electrical engineering.

Then there are loops (which we will consider in Chapter Four). Early in control theory, feedback loops were considered impossible, in logic, self-referential statements were considered illogical, and in decision theory, intransitive preferences are still considered irrational."

"So, in light of the above, we face this philosophical question: 'Does hyperdimensional space have to correspond to the space-time continuum of our universe in order to be useful for the understanding of mathematical models?' I claim the answer is 'no' and most mathematicians use as many dimensions as they need. Descartes' crucial intellectual leap was to enrich the algebraic relations with geometric concepts; the extrapolation to any number of dimensions should be trusted as a straightforward extension."

"OK," agreed the manager, feeling a little over-answered. "Your use of the projection and extension operators was less familiar to me and I never really thought how families of submodels contributed to the total model. The concept that a computational request is really a projection of the total model onto a subspace was really beyond my experience. But now I can see the value of this construct. The projection operator provides the dimensionally-limited human an understandable perspective of an inconceivable multi-dimensional relation."

"Or to further extend the lingo that managers like to use," added the analyst, "if you choose the right subdimensional viewspace, you get the 'best angle' on a complex problem – something you guys are always trying to do."

"Figure 2-14 is a good summary of most of the previous ideas. Referring to the ancient stories at the beginning of this chapter, the blind men each observing the elephant from different aspects form pitifully incomplete shards of truth, generating relations which are combined into the 'total truth.' However, since we are hopelessly subdimensional, we cannot perceive or understand the total truth, A. Instead, we turn it every which way and attempt to observe it from many different angles, some of which may help us to understand 'more deeply.' Disappointingly, for most of the directions we attempt to look at the relation, we will get no more information. That is the agony of asking questions which are not well-posed."

"*Bottom line*, to use more management jargon," summarized the analyst, "these four views were deemed necessary by the author to understand the underlying foundations of models, computations and well-posedness, rather than rely on the rather opaque, algorithmic crankturning he had been taught in all his courses in mathematics."

Generating Relations

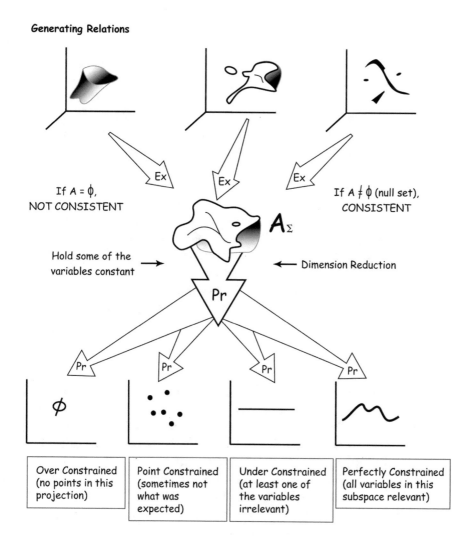

| Over Constrained (no points in this projection) | Point Constrained (sometimes not what was expected) | Under Constrained (at least one of the variables irrelevant) | Perfectly Constrained (all variables in this subspace relevant) |

Figure 2-14. In order for a problem to be well-posed, the mathematical model must be *consistent* and the computation must be *allowable*. *Consistency* requires that the hyper-dimensional relation not be the null set. *Allowability* requires that all the variables of the requested computation be relevant to the projection of the total relation onto the computational subspace.

2.8 CHAPTER SUMMARY

- Four interrelated views of a mathematical model and its computations have been presented: set theoretic, family of submodels, bipartite graph and constraint matrix. (Figure 2-15). The first two are full models, containing all the detail necessary for final construction and computation; the latter two are metamodels, and are abstractions of the first two which concentrate on the topological and computational features. In a strong sense, the metamodels can be considered to provide an overarching management perspective on consistency and computability issues. Without this perspective, those who attempt to build models and make computations on them will blunder into difficulties due to either inconsistencies in the model or unallowabililties in the computations – in short, the traditional well posed problems. In any case, once the "well-posedness" of the models and computations have been analyzed and managed by the metamodels, the full models must then be employed for the actual computations.
- The concept of "set" has been used as frequently in this chapter as the word "system" has been used in a book on systems engineering. This was done deliberately because – despite the apparent simplicity of the concept – it is far more precise a concept than "system" and its applicability is wide ranging.
- The *set* was used to define a *relation* between variables. The concept of *set* was also used to identify:
  - the allowable values of a variable
  - the possible values of a product set
  - collections of variables – which can represent computational requests
  - collections of relations – which can represent submodels
  - subsets of bipartite graph vertexes called knots
  - subsets of bipartite graph vertexes called nodes
  - collection of edges connecting subsets of the knots and nodes
  - constraint matrix columns; homomorphic to knots and variables
  - constraint matrix rows; homomorphic to nodes and relations
  - constraint matrix elements; homomorphic to edges and relevancies

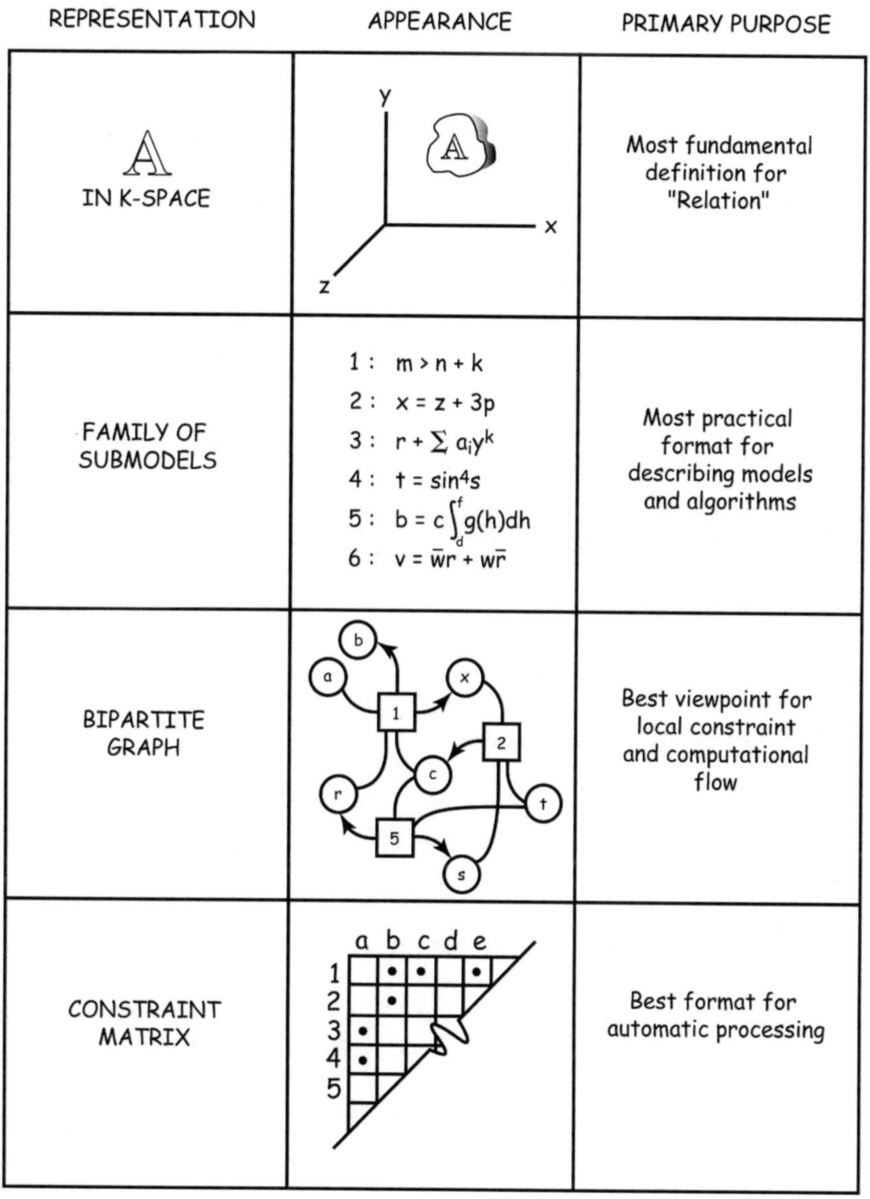

REPRESENTATION	APPEARANCE	PRIMARY PURPOSE
$\mathbb{A}$ IN K-SPACE		Most fundamental definition for "Relation"
FAMILY OF SUBMODELS	1 : $m > n + k$ 2 : $x = z + 3p$ 3 : $r + \sum a_i y^k$ 4 : $t = \sin^4 s$ 5 : $b = c \int_d^f g(h)dh$ 6 : $v = \bar{w}r + w\bar{r}$	Most practical format for describing models and algorithms
BIPARTITE GRAPH		Best viewpoint for local constraint and computational flow
CONSTRAINT MATRIX		Best format for automatic processing

Figure 2-15. The four representations of a mathematical model. The first two are full models and the latter two are metamodels.

- o power set of the knots: all definable computational requests
- o power set of the nodes: all possible submodels

- Perhaps Friedman's greatest contribution was the recognition that very useful metamodels of the mathematical model's variables, relations and relevancies are the bipartite graph's knots, nodes and edges and the companion matrix's rows, columns and elements.
- This chapter provided the foundation for a building; its construction and use will continue in subsequent chapters.

2.9 PROBLEMS FOR THE INTERESTED STUDENT

1. Provide a real-world example of a three-dimensional product set where one dimension is continuous, another is discrete and a third is defined in intervals.

2. Employing detailed algebraic equations in three-dimensional space, show an example of y being relevant to the model and another example of y being irrelevant.

3. For a three-dimensional model, show how ∩, ∪, Pr and Ex can be used to combine a family of three algebraic equations into the total model.

4. Draw the constraint matrix for Figures 1, 5, 6 and 7 of Chapter One.

5. Draw the constraint matrix for Figure 3 of Chapter One. Can you suggest how the term, "Basic Nodal Square" was developed?

6. Regarding the mathematical model depicted in Figure 1-3, which of the following computational requests are allowable and which are not allowable?

   For the allowable requests, draw the directed bipartite graph which depicts the computational flow. For the unallowable requests, discuss the reason(s) for the unallowability.

   Computational Requests:
   E=f(T,M), A=f(T,E), A=f(P,M)

Chapter 3 GENERAL RESULTS

From Protomath to Math to Metamath

3.1   LANGUAGE AND MATHEMATICS

"You've opened my mind to many new concepts and definitions, but I don't see where this is all leading," complained the manager. "Are there *results* I can use? I feel as if I'm learning the vocabulary of a new and rich language, but I can't make sentences."

"Your language/mathematics analogy is very apt," complimented the analyst. "Before the emergence of modern language humanity had probably millions of years of *protolanguage.* Words were formed to represent abstractions from the observed world and originally served to communicate basic ideas such as danger warnings or cooperation in game hunts. Full language appeared when the words were organized into complete-thought sentences employing grammar and syntax – which has a remarkable worldwide structural similarity over all known languages. Similarly, before the emergence of mathematics there were probably thousands of years of *protomathematics.* Concepts of number and geometry were employed in prehistoric times in practical ways for commerce and property surveys. Full mathematics appeared with the organization of "math facts" into a logical structure of definitions, relations and proofs – which also have a remarkable similarity across all cultures and languages."

"Although the two appear different superficially, mathematics is completely imbedded within language. All the rules of math and logic of proofs are linguistic. The applicability of language to the world is orders of magnitude greater than math; however when math is applicable, it offers these important advantages: precision, consistency, calculatability, generalizability, stability across cultures and languages, and perhaps most

© Springer International Publishing AG 2017
G.J. Friedman, P. Phan, *Constraint Theory*, IFSR International Series on Systems
Science and Engineering, DOI 10.1007/978-3-319-54792-3_3

importantly, provides a trustworthy vehicle to deduce conclusions from a great diversity of inputs. True, it can be argued that math treats only a tiny percentage of the world compared to language, but it was crucial to all the advances of science, technology and economics which contribute to our modern civilization."

"Now just as math employed higher order abstractions based in language, *metamathematics* employs abstractions of objects based in mathematics. Constraint Theory is a form of metamathematics which employs the bipartite graph and constraint matrix whose elements are the mathematical objects of variable and relation. Thus, constraint theory is yet one more step removed from the understanding we attain from natural language. That may be a disadvantage if it appears the mathematical complexity is formidable – but I argue the complexity is well below that of other branches of math. The advantage is that metamodels of this type actually can bring us *closer* to methods of plausible reasoning and further enable mathematics to intensify its beneficial augmentation of language in all the dimensions listed above."

"There is a fundamental mystery pondered by writers such as Devlin [13]: '*Why does language come so easily to virtually every human and why is math so hard – even terrifying?*' A typical child acquires a vocabulary of several thousand words and speaks in a respectable grammar even before formal education begins. Protomath is also acquired quite early. But education in math comes much later and never finds a comfortable place in the minds of most people. Devlin argues that the primary reason is that these people never live in what he calls the "math house" where the objects and logic of mathematics become as familiar as the everyday objects around which we form language. Even highly trained engineers who successfully apply integral transforms in the design of sophisticated control systems in their early careers are loath to trust math applied to model building or decision making as they take on management roles in their later careers. You, sir, are an excellent example of this math phobia in highly educated people."

The manager's glazed over eyes sharpened. "I would resent that remark, but I don't disagree enough. I've been exposed too many times to mathematicians claiming to serve me by narrowly focusing on a small part of the problem – the only part where math is applicable – and by attempting to impress by presenting their incomprehensible results within a forest of incomprehensible derivations. I've often wondered if the word 'analyst' had its roots in 'anal retentive.' Invariably, I had to come in to add the necessary additional dimensions and provide management judgment – whatever that means. Admittedly, when I did do math myself, I trusted that the proofs I was given were correct, and tended to skip over them as I was a student.

Even the formulas and rules which I applied were not always provided with the necessary assumptions to clarify their domains of applicability."

"A major advantage that math has over language is its *generizability* into domains which were previously incomprehensible," the analyst pontificated. As was mentioned in Chapter 1, we cannot really perceive numbers over seven or so – the way that we (and many animals) can perceive 1, 2 and 3. But we so trust the algorithms of arithmetic that we at least have a feeling or control of understanding *that which we need to know* about numbers into the millions or billions: 'which number is greater and by how much?' 'is number *a* slightly greater than *b* or is it an order of magnitude greater?' These useful answers can be obtained without fully perceiving large numbers. Similarly, the objective of constraint theory is to provide the manager of large models a trustworthy method to obtain answers to certain important properties of models: 'Is the model consistent?', 'Is the computation *a=f(b,z)* allowable on this model?' These answers are useful whether or not the manager can actually perceive the very high dimensions involved."

"OK, thanks for the sermon," said the manager, feeling that the explanations were somewhat long. "Let's see how you can bring me into Devlin's "math house" and what trustworthy results you have to provide."

3.2 MOST GENERAL TRUSTWORTHY RESULTS

__Theorem 1__: If a model is inconsistent, then no computational requests on it are allowable.

Proof: By Definition 13, the relation set of an inconsistent model is the null set. By Definition 14, for a computational request to be allowable, the projection of the model relation set onto the view space must have at least one point. But the projection of a null set onto *any* view space must also be the null set. Thus, if the model is inconsistent, any request is unallowable. *QED*

"Is the reverse also true?" asked the manager. "Depends on what you mean by 'reverse'," responded the analyst. "If you mean, 'does noncomputabililty imply inconsistency?,' the answer is 'no' – there are other reasons for noncomputability than inconsistency; refer to Appendix A. If you mean, 'does consistency imply computability?' the answer is 'no' again for the same reason. But if you mean, "does computability imply consistency,' then the answer is 'yes'. We have to be careful as to how we employ double negatives; not all natural languages recognize a double negative as a positive. A mathematical implication which is two-way is called 'iff' – if and only if." See Appendix C for a more thorough discussion of this logic. But I fear we are digressing from the main thread here.

"I couldn't fail to disagree with you less," commented the manager, self-referentially.

Theorem 2*:* If any submodel of a total model is inconsistent, then the entire model is inconsistent.

Proof*:* The total model relation set is the intersection of the relation sets of all its submodels. Since the intersection of any set with the null set is the null set, if any submodel relation is the null set, then the total model will also become the null set and thus, by Definition 13, will be inconsistent. *QED.*

"These are a good start it seems," commented the manager, but aren't they rather intuitive? And thus perhaps not so useful?"

"Well I hope all the remaining theorems will be as intuitive, or at least as plausible as these," the analyst responded. "As far as useful goes, it tells us to check the consistency of a model *before* we attempt computations on it. In my experience, most of the time computations fail, the model is checked only *after* much wasted effort. The utility goes further: even if a part of the model that is not used in the computation is inconsistent, then the computation is not allowable. Look at Figure 3-1. At first glance, it would appear that the computational request *d=f(a,c)* is allowable. However, observe that on the left side of the model, we have over three relations constraining the two variables *e* and *f*; this is a serious case of overconstraint – that is, the relation set for the submodel containing *e* and *f* is the null set. Therefore, *no* computations, including *d=f(a,c),* are allowable."

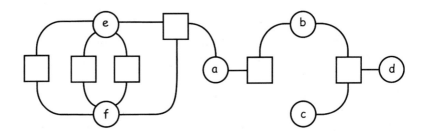

Figure 3-1. Inconsistency in one part of the model poisoning the whole model.

"OK, *that* seems nontrivial," admitted the manager. "I would have treated the *d=f(a,c)* computation as a work in progress and suppressed the remainder of the model as requiring repair. I note that even if the overconstrained submodel were reduced to just one basic nodal square, *a* would be intrinsically constrained, thereby not permitting *a* to be an independent variable."

"Excellent observation," complimented the analyst, "but you rarely know in advance all the computational requests you wish to make on a model and what portions of the model will be required for the request's computational paths. Thus, *any* submodel with a null set relation set can poison the entire model. An even more severe case is shown in Figure 3-2. Here the overconstrained submodel is in a completely separated component – there being no possible computational paths between it and the computational request – and the *total* model is still inconsistent, rendering all computational requests unallowable."

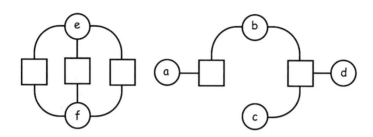

Figure 3-2. Inconsistency even in a disconnected component poisoning the whole model.

"Checking total model consistency seems to be a formidable task; every pair of relations should be examined to see if they produce a null set," worried the manager.

"It's nowhere near that bad," the analyst reassured, "look at Theorems 3 and 4."

Theorem 3: If two relations have no relevant variables in common, they are consistent with each other.

Proof: Assume two relations: relation 1 with relevant variables $a,b,c,..$ and relation 2 with relevant variables $r,s,t,...$. Choose any point in relation 1 – say $a_1,b_1,c_1...$ – and extend it into $r,s,t..$ space, resulting in the set defined by: $\{a_1,b_1,c_1....r,s,t...\}$. Similarly, choose any point in relation 2 – say $r_2,s_2,t_2...$ – and extend it into $a,b,c...$ space, resulting in the set defined by $\{a,b,c..r_2,s_2,t_2..\}$.

Now since the r,s,t of the relation 1 extension can take on any value, set them equal to the r_2,s_2,t_2 of the relation 2 extension, and set the a,b,c of the relation 2 extension equal to the a_1,b_1,c_1 of the relation 1 extension. Thus the two sets of coordinates are identical and we have guaranteed that there is at least one point in the intersection of the extensions of the two relations. *QED*. (See Figure 3-3 for a simple example of this process.)

"This substantially eases the task implied by Theorem 2; only relations which are 'adjacent' – that is, have relevant variables in common – can engender inconsistency."

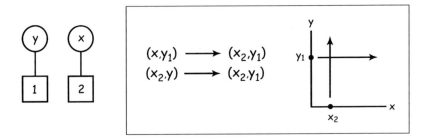

Figure 3-3. A simple example of the consistency of relations without common relevant variables.

Definitions 15: If a pair of relations lie in subgraphs such that there exists at least one path of edges connecting them, they lie in a *connected component.* If the pair of relations lie in subgraphs with no path connecting them, they lie in *disconnected components.*

Theorem 4: If two disconnected components are internally consistent, they are consistent with each other.

Proof: Any two relations which lie in separate components cannot share variables in common. Therefore, by Theorem 3, they are consistent. *QED.*

Theorem 5: No computations involving variables from disconnected components are allowable.

Proof: For computational allowability, Definition 14 requires that the intersection of the projections of the relations not be the null set and that the request's variables be relevant to it. The first requirement is satisfied but – because there are no relevant variables across disconnected sets – the second requirement is not met. Thus all computations across disconnected sets are not allowable. *QED.*

"Now *that's* certainly plausible," said the manager. "I wouldn't expect that one could compute across disconnected components which are really islands in separate universes."

"Agreed," agreed the analyst.

Theorem 6: The allowability of a computational request is independent of permutations of its dependent and independent variables.

Proof: An allowable computational request will have a satisfactory relation set in the sense of Figure 2.14. Choosing any dependent variable out of the request's variables by merely rotating the view space will not alter the validity of the relation set. *QED.*

Theorem 7: All possible computational requests on a model with K knots (variables) is the power set of the knots and its number is equal to 2^K.

Proof: Each computational request is a subset of the set of knots and can be uniquely identified with a binary number whose length equals the number

of knots. Therefore the number of subsets in the set of K knots equals 2^K.
QED. Refer to Figure 3-4 for demonstrations and simple examples.

Assume that the set N contains 2 elements: {a,b}

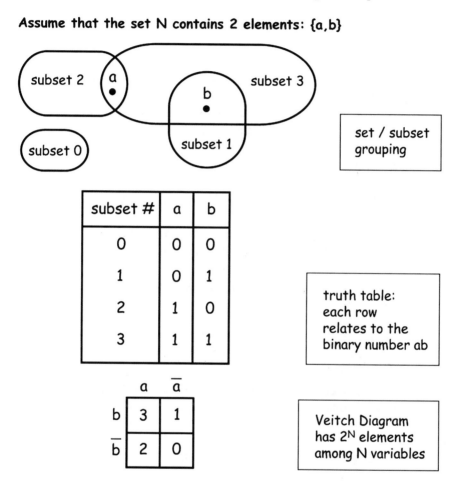

Figure 3-4. Three demonstrations that the power set of the set of N elements contains 2^N subsets.

"If it weren't for Theorem 6," commented the manager, "the number of possible computational requests predicted by Theorem 7 would be far larger, I presume."

"Certainly," agreed the analyst, "for each of the 2^K subsets of dimension *d*, one could choose *d* dependent variables, enormously increasing the possible computational requests. Of course, one could still choose different ways to *plot* the results of the computational requests, but that is really outside the concern of constraint theory, which is mainly concerned about the fundamental allowability of any of the plots."

Theorem 8: All possible submodels of a model with N nodes (relations) is the power set of N and its number equals 2^N.

Proof: This proof is identical to that of Theorem 7; merely replace K with N.

Definition 16: A *tree* is that structure within a connected component of a graph such that there is exactly one path connecting every pair of vertices.

Definition 17: A *circuit cluster* is that structure within a connected component of a graph such that there are two or more independent paths connecting every pair of vertices. *Independent paths* share only their initial and terminal vertices. *Adjacent circuits* are circuits which share at least one edge.

Refer to Figure 3-5 for examples of trees, circuits, adjacent circuits and circuit clusters.

Definition 18: A *universal relation* is a relation which does not limit any of its relevant variables to a given range. For example, $x+y=5$ and $m=n^3$ are universal relations, but $s^2+t^2=3$ and $z>4$ are not.

Theorem 9: Any set of universal relations whose bipartite graph has a tree structure is consistent.

Proof: First, prove that any two universal relations that have only one relevant variable in common are consistent. (Inconsistency could occur if the common variable had incompatible constraints placed on it by the two relations, but all the variables, by Definition 18, have unlimited ranges.)

Then append additional relations to the model in each case with only one variable in common, forming a tree. Consistency will be maintained at each step. *QED.*

3.3 CLASSES OF RELATIONS

Thus far, all the results and discussions of consistency and computability have been on the basis of any conceivable type of general relation. As was discussed in Chapter 2, relations can take on an extremely wide variety of properties. In order to progress towards effective tools for the management of multidimensional math models, it will be necessary to define three important classes of relations.

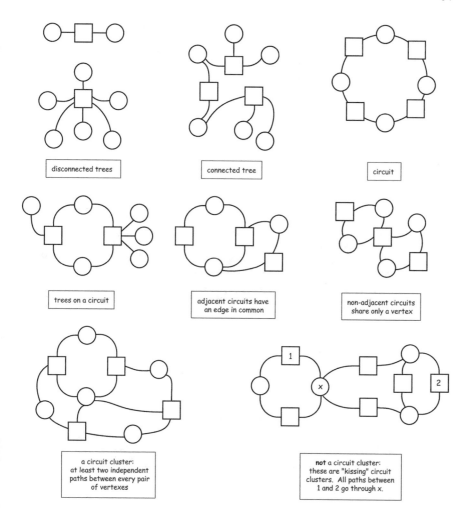

Figure 3-5. Examples of trees, circuits, adjacent circuits and circuit clusters.

Definition 19: *Relation classes:* Consider the general relation A_2. Let l be a line through any point in A_2. Let $A_2 \cap l$ be the intersection of A_2 with l, and let $Pr_l A_2$ be the projection of A_2 onto l. Then:
  - A_2 is a *discrete relation* if both $A_2 \cap l$ and $Pr_l A_2$ are point sets of measure zero.

- A_2 is a *continuum relation* if $A_2 \cap 1$ is a point set and Pr_1A_2 is an interval set of non-zero measure.
- A_2 is an *interval relation* if both $A_2 \cap 1$ and Pr_1A_2 are interval sets of non-zero measure.

Examples of these three relation classes are given in Figures 3-6 and 3-6A.

Discrete Relations

1: $a = b\bar{c}d + \bar{c}\bar{e}f + c\bar{g}f$

2:

y ↓ \ x →	1	3	5	7	11	13	17	19	23	29	31
2		•				•			•		
4			•			•			•	•	•
6	•		•		•		•		•		

3: $\sum_{n=0}^{N} a_n x^n = 0$

Continuum Relations

4: $m = \sum_{p=1}^{n} np^n$

5: $r = s^x + \int_0^\infty f(t) \exp(-st)\, dt$

Interval Relations

6: $x^2 + 2y > 3$

7: $Pr_V[U^2 + 3V^2 = 1]$

Figure 3-6. Examples of the three relation classes.

These three relation classes will be treated in the next two chapters. We will discuss the continuum relation class first – in Chapter 4 – because it can be argued that it is the most important for math model building and also because constraint theory happens to be most useful for this class. Chapter 6 will treat discrete and interval relations.

Relation Classes

CLASS NAME	DISCRETE	CONTINUUM	INTERVAL
PICTURE	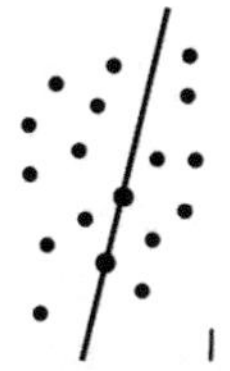		
NUMBER OF POINTS IN A ∩ 1	finite	finite	infinite
NUMBER OF POINTS IN Pr₁A	finite	infinite	infinite

Figure 3-6A. Three relation classes can be defined – at least locally – by the number of points in their intersections with a line and the number of points in their projections onto a line.

3.4   MANAGER AND ANALYST REVISITED

"I'm beginning to see what you mean by 'living in a math house,'" the manager complained mildly. "Although all the definitions and Theorem proofs are imbedded in natural language, they appear far 'tighter and rigorous' than the ambiguities of normal conversation. It certainly doesn't make for fast reading. Especially if one wants to understand the proofs."

"I'm trying to make it as plausible and painless as I can," responded the analyst. "I encourage you to read the proofs and get in the spirit with the mathematical rhythm of this theory. Otherwise, you won't gain as much confidence in the trustworthiness of extending these results into the high dimensions that we need in order to manage modern models. So far, you've just entered the foyer of the 'math house'; I want to show you at least three more rooms."

3.5   CHAPTER SUMMARY

- Language and mathematics are two of humanity's greatest gifts and are remarkably similar. However mathematics is more precise and is embedded within language which is more general. We need to become familiar with 'the math house' in order to extend our results to unimaginably high dimensions.
- Model consistency is a necessary requirement for computational allowability. If any part of a model is inconsistent, the entire model is therefore inconsistent, but relations with no common relevant variables are consistent. Universal relations with a tree graph structure will always be consistent. Computations across disconnected components are not computable.
- All possible computational requests are the power set of the set of the variables and they number 2^K; all possible submodels are the power set of the relations and they number 2^N.
- In order to provide a basis for more specific results, three classes of relations are defined:
    - *discrete*, dealing mainly with point sets such as Boolean logic;
    - *continuum*, dealing mainly with continuous curves; and
    - *interval*, dealing mainly with densely packed sets such as x>5.

3.6   PROBLEMS FOR THE GENERAL STUDENT

1. Construct a simple example for Theorem 1 showing that consistency is necessary for computability.
2. In the example given in chapter 1, was the model provided consistent? If so, why weren't all the computational requests allowable? If not, why were some of the requests allowable?
3. Construct a simple example for Theorem 2 showing that any submodel inconsistency "poisons" the entire model.
4. Given a model with K=4, show all the possible computational requests and show that they number 2^K. (For the sake of completeness, the full set and the empty set are considered valid "subsets".)
5. Construct a simple example demonstrating the validity of Theorem 9.

Chapter 4 REGULAR RELATIONS

Searching for the Kernels of Constraint

4.1   COGNITIVE BARRIERS TO CIRCUITS

"I must admit that the foyer of the math house was fascinating," said the manager, "and the rigorous structure based on the previous definitions was quite impressive. With the theorems and their proofs, I guess I've progressed from protomath to full math. However, it was a little like the ten commandments: after an admonition on how much I should respect and revere this central philosophy, all I got was a series of *negative* statements: can't kill, can't commit adultery, can't compute if inconsistent... I'm still looking for useful rules which will permit me to *manage,* as this book's title promises."

"This Chapter will present many rules and procedures by which you can more effectively manage large math models," assured the analyst. "In fact, it will end with a *Constraint Theory Toolkit* which summarizes the most useful of the theorems, rules and procedures."

"Central to many of the rules will be the treatment of circuits and loops within extremely tangled bipartite graphs. Many researchers appear to have a basic phobia about circuits. Logicians dislike self-referential loops because of the potential for paradoxes. Early in the field of control systems, feedback was thought to be illogical. Even von Neumann classified intransitive loops of preferences as 'irrational.' (See Figure 4-1) Devlin [12] noted that the common structure of language worldwide was treelike and hypothesized that 'our cognitive wiring favors trees.' So I'm asking you to keep an open mind on the concept and value of circuits and loops."

© Springer International Publishing AG 2017
G.J. Friedman, P. Phan, *Constraint Theory*, IFSR International Series on Systems Science and Engineering, DOI 10.1007/978-3-319-54792-3_4

Circuitphobia	
PRECEIVED PROBLEM	**POSSIBLE OUTCOME**
Non-sequential computation	Solution of simultaneous equations
Feedback	Control stability
Non-tree reporting structures	Horizontal management
Self referential logic	"consciousness"
Self replication	reproduction in automata
Auto-catalysis	Origin of "life"

Figure 4-1. The concept of circuits has initially appeared to be contrary to rational thought, but their careful management has led to many advances.

"I'll try, but this is the briefest philosophical introduction you've given to any chapter so far," chided the manager. "I've attempted to enjoy your pontifications on cognitive science, language, origins of mathematics and the universe. Are we properly warmed up for the meat of this chapter?"

"Yes, I hope so," responded the analyst. "We have more material in this chapter than any other. It is the heart of Constraint Theory so far in its development. I'm anxious to get going."

4.2 NODE, KNOT AND BASIC NODAL SQUARE SANCTIFICATION

***Definition 20*:** A pair of relations are *locally universal* if the ranges and domains of their relevant variables are mutually compatible. In other words, any output from one relation is an acceptable input to the other. For example the circle $x^2+y^2=1$ is locally universal with x=0, but not with x=3.

Postulate 1: Model builders inherently wish their relations to be locally universal.

Although an important assumption for constraint theory analysis, the actual testing of universality can only be accomplished on the full model; the constraint theory metamodels do not have sufficient information.

Definition 21: A set of *regular* relations are continuum relations which are locally universal with all their interacting relations.

Definitions 22: The *constraint potential* of a graph G is defined as p(G) and equals the excess of nodes over knots: p(G)=N-K. It is the negative of the "degrees of freedom" notion used in Chapter 1. It will be useful to define two circumstances where constraint potential is insightful:

> p_i(G)=*intrinsic constraint potential;* prior to any computational request,

> p_r(G)=*resultant constraint potential;:* after the application of independent variables, constants and computational flow from neighboring portions of the bipartite graph.

Definition 23: The *degree of a vertex, d(v),* equals the number of arcs that intersect that vertex. For the node vertex, d(n), it is the number of variables that are relevant to the relation; for the knot vertex, d(k), it is the number of relations which employ that variable. Refer to Figure 4-2 for examples.

Thus, for example, the constraint potential of a node with degree d(n) is merely 1-d(v) since it represents one node attached to d(n) knots. In general, the average degree of all of a graph's vertices is a strong indication of the graph's connectivity.

Theorem 10: For a model graph of regular relations, with a tree-like topological structure, the computational rules are:

for nodes: d(n)-1 inputs permit 1 output;

for knots: 1 input permits d(k)-1 outputs.

Proof: Consider the d(n)-dimensional space of any given node: if d(n)-1 of these dimensions are chosen as inputs, then this will define a line which will intersect the relation in a point (or set of points), by definition 19, since the relation was assumed to be regular and thus a continuum relation. QED_1.

The single output of each of these computations can then propagate to all other d(k)-1 nodes for which this variable is relevant and, by Definition 21, will be compatible with all of them because a regular relation is locally universal. QED_2.

Thus the intuitively appealing rules which were employed so extensively in Chapter 1 are now rigorously "sanctified" by Theorem 10. This concept will be generalized to all topological structures later in the chapter.

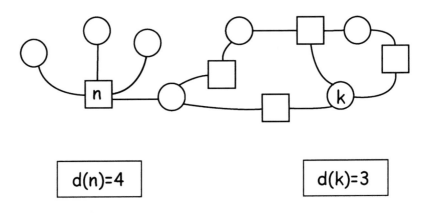

d(n)=4   d(k)=3

Figure 4-2. The degree of a vertex is merely the number of edges which intersect it.

Definitions 24: If any node on the path has a resultant constraint potential greater than zero, the computational request is not allowable due to *overconstraint*. If any node on the path has a resultant constraint potential less than zero, the computational request is not allowable due to *underconstraint*. If the entire path has a resultant constraint potential of zero, the computational request is *perfectly constrained* and is allowable. Refer to Figure 4-3.

However, most math models – even those which could be called very loosely connected – will have bipartite graphs with circuit structures. When the computational flow described by the above rules reaches the vicinity of a circuit, over- and underconstraint cannot always be determined unambiguously, and a new rule will be required. For this purpose, it is necessary to define a special type of structure within the bipartite graph and constraint matrix:

Definitions 25: A *Nodal Square, NS,* is a submodel of a math model such that its constraint potential, p(NS)=0. A *Basic Nodal Square, BNS,* is a nodal square which does not contain a smaller nodal square within it.

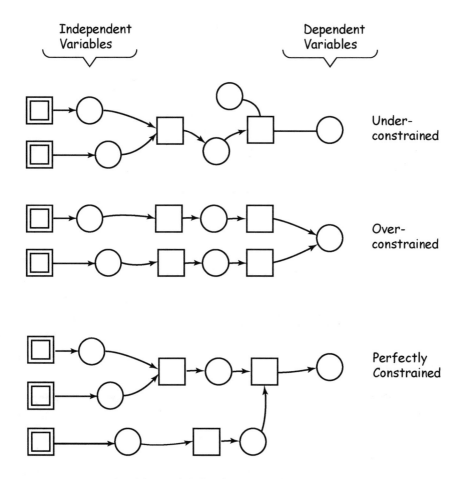

Figure 4-3. Computational (constraint) flow in trees requires only the simple rules: "d(n)-1 in, 1 out for nodes; and 1 in, d(k)-1 out for knots."

Fundamentally, nodal squares and basic nodal squares have the property that, in their local submodel, the number of variables equal the number of relations. Examples are shown in Figure 4-4. Recall that submodels are formed from the total model by grouping subsets of the *nodes* (or rows of the constraint matrix); thus the term, "nodal" squares if all the elements of the constraint matrix are captured in a "square." In this context, groupings of the knots (or columns of the constraint matrix) have no meaning. As defined earlier, groupings of the variables form the power set of all possible computational requests.

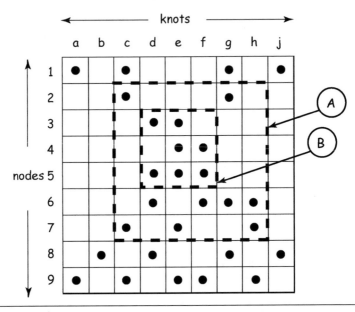

A: The submodel {234567} × {cdefgh} is an NS but not a BNS because of B

B: The submodel {345} × {def} is a BNS contained within an NS

REMEMBER:

A submodel is a grouping of nodes (with their relevant variables), NOT a grouping of knots.

Figure 4-4. A *Nodal Square*, NS, is a grouping of the rows of a constraint matrix, [C], such that all the relevant elements are captured within a square. That is, the submodel has an equal number of relations and variables. A *Basic Nodal Square*, BNS, is an NS which does not have a smaller NS within it.

This brings us to the second computational theorem of this chapter:

Theorem 11: Every Basic Nodal Square (BNS) of regular relations exerts point constraint on each of its relevant variables. That is, all its variables are constrained to either points or sets of points.

Proof: First, note the "dimension reducing" property of regular relations. A regular relation applied to k-dimensional space will form an allowability set of dimension k-1. If a second regular relation is applied, the intersection of the two allowability sets will have dimension k-2. In general, if n sets are intersected, the resulting dimension will be k-n. (Refer to Figure 4-5) Since the BNS has by definition a constraint potential of zero, K-N=0 and thus the intersection of all the N relations has a dimensionality of zero. *QED.*

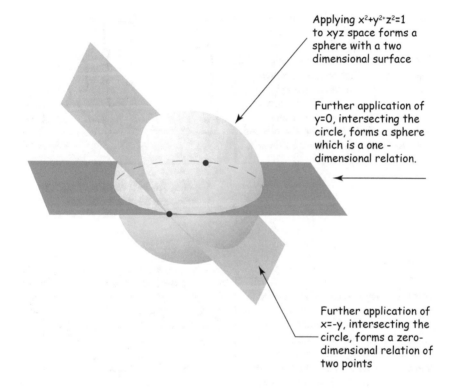

Applying $x^2+y^2+z^2=1$ to xyz space forms a sphere with a two dimensional surface

Further application of y=0, intersecting the circle, forms a sphere which is a one - dimensional relation.

Further application of x=-y, intersecting the circle, forms a zero-dimensional relation of two points

Figure 4-5. Regular relations have the property that each application of a new relation reduces the dimensionality of the total relation by one.

"I have no argument with this result," commented the manager. "Theorem 11 merely tells us that if we have *n* simultaneous equations involving *n* variables, then we should be able to solve for all these variables. In fact, I observe that the BNS can even be a 1x1 square: one equation and one unknown, which I also expect to always be able to solve."

"Yes, I agree that it is intuitively appealing," responded the analyst, "but that intuition is based on the career-long experience of mindlessly manipulating algebraic rules. Theorem 11 "sanctifies" this algebraic rule from the broader perspective of multidimensional relation theory. Perhaps what is less intuitive – and therefore more valuable – is that the BNS is the "kernel of constraint" in multidimensional math models. Figure 4-6 shows why the nodal square (NS) is not the kernel of constraint. In some cases the 'shell' between the NS and BNS is merely the resultant constraint domain emanating from the BNS' sources of constraint; in other (more serious) cases, part of the NS is overconstrained with overlapping BNSs and the remaining part is tree-like and underconstrained."

This bipartite graph is a Nodal Square, NS, since the number of relations and variables are equal. However, the kernel of constraint is the Basic Nodal Square, BNS, 12ab within it, which constrains the variables a and b. In the remainder of the NS, constraint merely flows from knot b to knot c. Thus, the "shell" between the NS and BNS is a resultant constraint domain

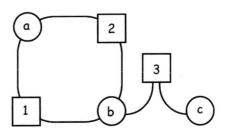

(a) An NS with a resultant constraint domain

This bipartite graph is also a Nodal Square since K=N. However, it contains three BNS's which overconstrain the variables w and x. Even if x were not overconstrained, there is still no flow from knot x to either knot y or z. Thus in this NS, not all the knots are even constrained at all, and the kernel of constraint is still the basic nodal square.

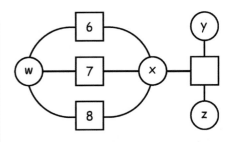

(b) An NS which both over- and underconstrains its knots

Figure 4-6. The Nodal Square (NS) is not the kernel of constraint; only the Basic Nodal Square (BNS) is. In NS-type a, constraint flows from the BNS to fill out the NS. In NS-type b, multiple BNSs overconstrain some of the knots and constraint doesn't even reach the other knots.

"Thinking back to the simple example of Chapter 1, let's examine the general propagation of constraint across a math model," suggested the analyst, referring to Figure 4-7:

Even before a computational request is made, we must determine if the model is consistent. As we observed earlier, there may be *intrinsic sources of constraint* in the form of BNSs (including 1x1 BNSs) that either point constrain the model or, worse yet, overconstrain it if the BNSs overlap common variables. If the BNSs overlap and contribute to overconstraint, then the model is not consistent and we can go no further until the overconstraint is relieved. Worse yet, the constraint may propagate into *resultant constraint domains,* possibly overconstraining variables which are in two or more of these resultant domains. Again, if this occurs, the model is not consistent; overconstraint must be relieved before we can ask for computational requests.

When checking for computational allowability, the procedure is quite similar, except that, in addition to the constraint sources, we now must superimpose constraints in the form of independent variables and variables held at some selected constant value. As is shown in Appendix A, the application of the rules of Theorems 10 and 11 will generally not yield an allowable computational request. In tree structures, the Theorem 10 rules can be applied very rapidly, but they will likely bog down in the vicinity of circuit structures where the BNSs are hiding.

Therefore, for both the determination of model consistency and computational allowability, the location of the BNSs become critical; they are the "kernels of constraint" around which inconsistency and unallowability occur.

"Well," observed the manager, "they weren't hard to find in the example of Chapter 1. And for higher dimensioned models, all I need to do is merely examine the subsets of the constraint matrix rows to see where the BNSs are hiding."

"There's that word 'merely' again," chided the analyst. Appendix A deals with sizable models of thousands of variables and relations. However let's look at just a medium size model of say, 100 dimensions. The number of possible submodels – or subsets of the rows of the constraint matrix, as you put it – is the power set of the set of nodes and is equal to 2^{100}. Even if your computer could examine one of these subsets for a possible BNS every nanosecond, it would still take about 10^{14} years, or a *thousand lifetimes of the universe* to go through this power set exhaustively."

"Attempting to find BNSs by staring at the bipartite graph would be even worse," guessed the manager. "Hundreds and thousands of vertices would appear as a monstrous 'snake chart.'"

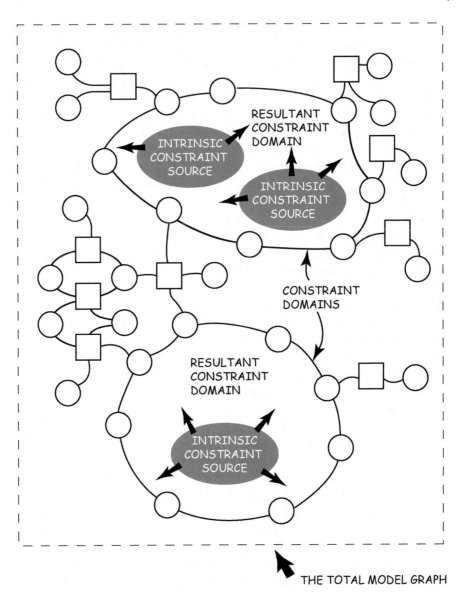

Figure 4-7. Even before a computational request is made, intrinsic constraint sources –
formed on the BNS kernels – may exist in consistent models. These sources flow their
constraint outwards into resultant constraint domains. If these domains overlap, overconstraint
of the variables within the overlap is likely, rendering the overall model inconsistent.

"Absolutely," agreed the analyst. "The low dimensional bipartite graphs
are certainly useful to develop general theories, but they would be hopeless
to use as a tool for realistically complex models. Figure 4-8 is an example of

how messy it can get. There is an important theorem in set theory by Hall [14] which we modify into bipartite graph theory as:

Theorem 12: In order that a distinct output variable be associated with each of *m* relations it is sufficient that, for each K=1,2,3..*m*, any selection of K of the relations shall contain between them at least K relevant variables.

The computer aided method is the only one feasible and we need the constraint matrix for communication with the computer. However, if we are to address realistic dimensions of modern models, we must avoid the trap of exhaustively searching for the BNS culprits through power sets with their attendant exponential explosions.

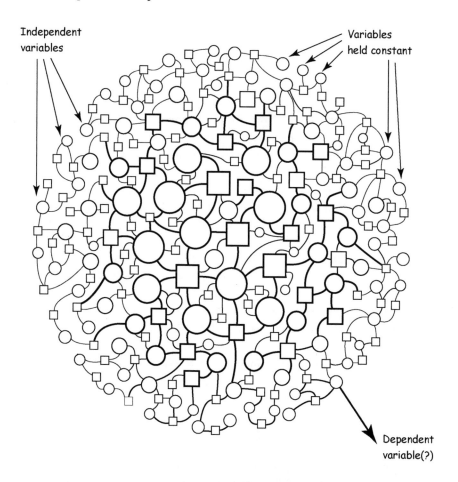

Independent variables

Variables held constant

Dependent variable(?)

Figure 4-8. A bipartite graph so large that gravity forces it into a spherical shape.

Our strategy will be to understand how the properties of the BNS fits within the easily computed properties of the bipartite graph. If we are alert and understand what we're doing, we can improve the computation time to find the BNS foxes in the dense forests of the bipartite graph by factors of trillions. But first, in Section 4.3 we must summarize some properties of the bipartite graph, before we develop practical rules to locate the BNSs in section 4.4.

4.3 USEFUL PROPERTIES OF BIPARTITE GRAPHS

The most obvious properties of any graph are its connectedness, its tree-ness and its circuit-ness. Each of these has important consistency and computability consequences and we will treat them in this order.

In order to focus on the structure of graphs, let us repeat these three important definitions from Chapter 3:

Definitions 15: If a pair of relations lie in subgraphs such that there exists at least one path of edges connecting them, they lie in a *connected component*. If the pair of relations lie in subgraphs with no path connecting them, they lie in *disconnected components*.

Definition 16: A *tree* is that structure within a connected component of a graph such that there is exactly one path connecting every pair of vertices.

Definition 17: A *circuit cluster* is that structure within a connected component such that there are two or more independent paths connecting every pair of vertices. *Independent paths* have no vertices in common except their end points.

Examples of definitions 15, 16 and 17 are provided in Figure 3-5. Let us first address an automatic algorithm to determine the connected components of a graph.

Definition 26: A node and knot are *adjacent* when there is an edge connecting them. A vertex is a *separating vertex* if its removal disconnects the graph.

From the viewpoint of the constraint matrix, the existence of a relevant element in the xth row and yth column denotes the existence of the xy edge and thus the adjacency of the x node and y knot.

Definition 27: The *connectedness algorithm*, by repeatedly applying the adjacency definition, will partition any graph into disjoint connected components.

Definition 28: The *separating vertex algorithm*, by employing trial eliminations of vertices and determining whether the remaining graph is still connected, will locate the graph's separating vertices.

These algorithms are exhaustive and will always work to determine the connected components and separating vertices of any graph, including bipartite graphs. However, it may sometimes be useful to employ more global results which are a function of only the total number of vertices and edges and provide faster conclusions on some occasions. These results, showing when the graph *must* be connected, *may* be connected, and *cannot* be connected are presented in Figure 4-9.

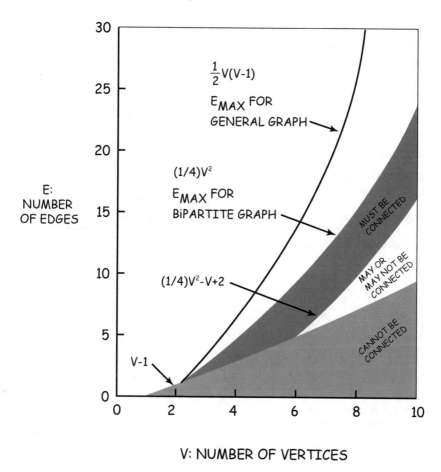

V: NUMBER OF VERTICES

Figure 4-9. Domains of connectedness for bipartite graphs as a function of the number of edges, E, and the number of vertices, V. (Recall that V=K+N.)

___Theorem 13:___ A connected graph which has V-E=1 is a tree. A tree is a minimally connected component and every vertex of a tree is a separating vertex.

___Proof:___ Consider the simplest possible case for a tree: two vertices connected by one edge; thus V=2 and E=1, yielding V-E=1. Now add more fragments which are connected but do not contribute more paths between vertex pairs. In all cases, these fragments will add one more vertex and one more edge; thus V-E=1 is still true no matter how many more fragments are added. *QED₁*.

Since a tree provides only a single path between any vertex pair, removing any vertex will interrupt that path and disconnect the tree into separate components. *QED₂*.

Theorem 13 provides us with a simple and powerful tool for the analysis of trees within graphs, no matter how highly dimensional and complex they become. Figure 4-10 presents examples of how it can be applied.

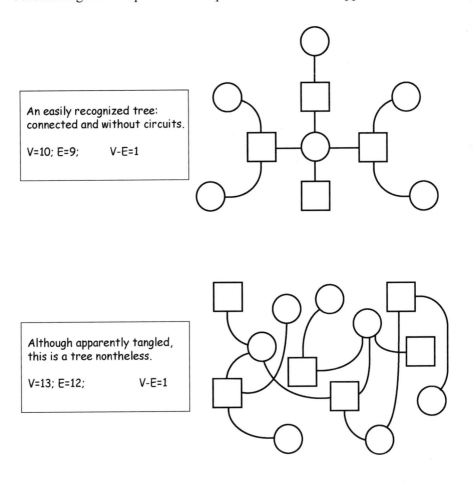

An easily recognized tree: connected and without circuits.

V=10; E=9;   V-E=1

Although apparently tangled, this is a tree nontheless.

V=13; E=12;   V-E=1

Figure 4-10. Theorem 13 is unerring in its ability to identify trees. Initial layouts of bipartite graphs may frequently disguise the fundamental "tree-ness" of the structure, especially when the dimension is large.

__Definition 29:__ The *tree algorithm for graph G with V vertices and E edges:* If V-E=1, then G is a tree; if V-E>1, then G is disconnected, if V-E<1, then G has at least one circuit.

A companion algorithm which is even simpler and in many cases can locate the very ends – or "twigs" – of the trees easily, is:

Definition 30: The *tree-twig pruning algorithm:* search the entire constraint matrix for vertices of degree one – solitary elements in a row or column – and remove the row (which is the node or relation). Repeat until no vertices of degree one remain. The residue will be a graph completely comprised of circuits, which may be connected by "internal trees" (without twigs).

Finally, let us consider the most complex case of circuits within graphs:

Definition 31: The *circuit rank, c(G),* of a graph, G, with V vertices, E edges and P connected components is: $c(G)=E-V+P$.

Definition 32: A *simple circuit,* C_j is a directed sequence of connected edges that does not use any vertex or edge twice. A *circuit vector* for circuit C_j is a sequence of elements, $\{e_1,e_2,e_3...\}$ where:

$$e_i = \begin{cases} +1, & \text{if } C_j \text{ traverses edge } i \text{ in positive sense} \\ -1, & \text{if } C_j \text{ traverses edge } i \text{ in negative sense} \\ 0, & \text{if } C_j \text{ does not traverse edge } i \end{cases}$$

For the purpose of determining the e_i, the edges of G are arbitrarily assigned directions.

Definition 33: *Circuit Vector Addition and Independence* have exactly the same meaning as vector addition and independence in linear algebra.

Refer to Appendix D for a brief summary of the relevant portions of linear algebra which we will use for graph theory analysis. This relationship between vector spaces and graph theory can be viewed as another example of the unity and even beauty of mathematics. Its utility is demonstrated by the following simple and powerful theorem:

Theorem 14: The number of simple, independent circuits of a graph G, equals $c(G)$, its circuit rank.

Proof: See Reference [2].

This theorem is truly amazing, providing us with very useful information about the complexity of circuit clusters, requiring only a knowledge of the number of the graph's vertices and edges – these are merely the semiperimeter and the number of non-trivial elements of the constraint matrix. The theorem essentially tells us the dimensionality of the circuit vector space. In linear algebra by comparison, there is no comparable method to compute the dimensionality of the vector basis of a space of many vectors.

A dramatic example of Theorem 14's application is shown in Figure 4-11.

A relatively simple circuit cluster has 13 simple circuits, but only 4 of them are independent circuits in the sense of Definition 30. This is analogous to a circumstance where 13 vectors are defined but they can be captured within a 4-dimensional vector space. In graph theory we can compute the 4 dimensions by a trivial computation of the observable variables V and E; but in linear algebra, the computation is far more arduous and becomes rapidly worse with higher dimensions. Figure 4-10a provides examples of circuit rank, $c(G)$, as well as constraint potential, $p(G)$.

Circuit Rank = c(G) = E-V+1;   Constraint Potential = p(G) = N-K

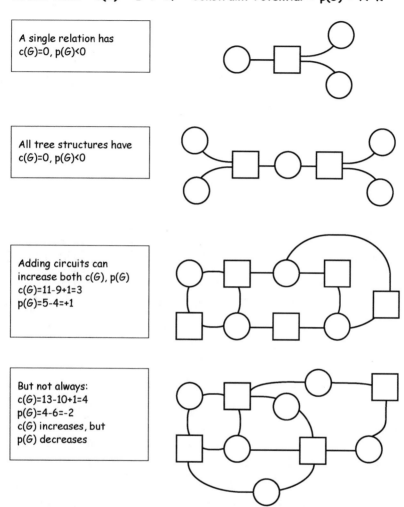

A single relation has
c(G)=0, p(G)<0

All tree structures have
c(G)=0, p(G)<0

Adding circuits can
increase both c(G), p(G)
c(G)=11-9+1=3
p(G)=5-4=+1

But not always:
c(G)=13-10+1=4
p(G)=4-6=-2
c(G) increases, but
p(G) decreases

Figure 4-10a. Both the circuit rank and the constraint potential of a bipartite graph will be useful to locate the kernels of constraint in a complex model.

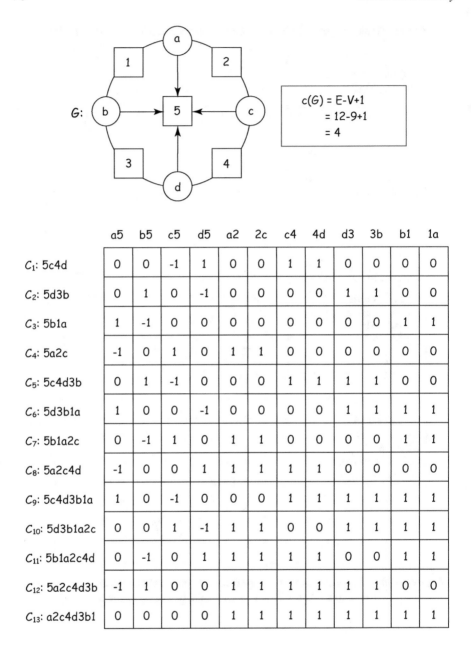

	a5	b5	c5	d5	a2	2c	c4	4d	d3	3b	b1	1a
C_1: 5c4d	0	0	-1	1	0	0	1	1	0	0	0	0
C_2: 5d3b	0	1	0	-1	0	0	0	0	1	1	0	0
C_3: 5b1a	1	-1	0	0	0	0	0	0	0	0	1	1
C_4: 5a2c	-1	0	1	0	1	1	0	0	0	0	0	0
C_5: 5c4d3b	0	1	-1	0	0	0	1	1	1	1	0	0
C_6: 5d3b1a	1	0	0	-1	0	0	0	0	1	1	1	1
C_7: 5b1a2c	0	-1	1	0	1	1	0	0	0	0	1	1
C_8: 5a2c4d	-1	0	0	1	1	1	1	1	0	0	0	0
C_9: 5c4d3b1a	1	0	-1	0	0	0	1	1	1	1	1	1
C_{10}: 5d3b1a2c	0	0	1	-1	1	1	0	0	1	1	1	1
C_{11}: 5b1a2c4d	0	-1	0	1	1	1	1	1	0	0	1	1
C_{12}: 5a2c4d3b	-1	1	0	0	1	1	1	1	1	1	0	0
C_{13}: a2c4d3b1	0	0	0	0	1	1	1	1	1	1	1	1

Figure 4-11. A bipartite graph with 13 simple circuits. However, a simple computation of the circuit rank, c(G)=E-V+1 reveals that there are only four *independent* circuits.

Definition 34: *The circuit rank of G algorithm:* compute the number of independent circuits of G by: c(G)=E-V+P.

Definition 35: A *taxonomy of graph structures*. The taxonomy listed below classifies all possible graphs by defining these five properties in sequence: a) number of paths between vertices, b) number of independent paths, c) existence of trees, d) whether the trees are internal or external, and e) whether the trees connect to other trees or to circuit clusters.

1. Zero paths connecting vertex pairs: disconnected components
2. One path connecting vertex pairs: isolated trees
3. Many paths connecting vertex pairs: circuit/tree structures
    3.1. All paths are independent: circuit cluster
    3.2 Not all paths are independent: circuit/tree networks
        3.2.1 Circuit clusters w/o trees: "kissing" circuit clusters
        3.2.2 Circuit clusters with trees:
            3.2.2.1 Trees are external: circuit clusters with "twigs"
            3.2.2.2 Trees are internal: "doily" structures
                3.2.2.2.1 Trees-trees
                3.2.2.2.2   Trees-circuit clusters

These definitions are summarized in Figure 4-12, along with a Venn Diagram demonstrating the nested nature of the sequential categories.

Theorem 15: In the taxonomy of Definition 35, the following categories are mutually exclusive and exhaustive: 1, 2, 3.1, 3.2.1, 3.2.2.1, 3.2.2.2.1 and 3.2.2.2.2.

Proof: Examination of the nested category specifications – with the aid perhaps of Figure 4-12 – demonstrates that the specifications are mutually exclusive and exhaustive at every level of classification. Specifically, 3.2.2.2.1 and 3.2.2.2.2 make up 3.2.2.2, which together with 3.2.2.1 make up 3.2.2, which together with 3.2.1 make up 3.2, which together with 3.1 make up 3, which together with 1 and 2 make up all possibilities. *QED*.

In the next section, we apply the above properties of graphs in order to locate the BNS kernels of constraint in a far more efficient manner than the brute force approach mentioned above.

4.4   CORNERING THE CULPRIT KERNELS; TEN EASY PIECES

Now that we're somewhat familiar with some aspects of "the math house of graphs" we can extend the previous section's results on general graph theory to bipartite graphs and corner the BNSs which may be lurking deep within the tangled web of models like Figure 4-8.

NUMBER OF PATHS	NUMBER INDEP PATHS	WITH TREES?	TREES INT/EXT	TREES CONNECTED TO?	MEAE, SET
1. zero					*
2. one					*
3. many					
	3.1 all				*
	3.2 not all				
		3.2.1 w/o			*
		3.2.2 with			
			3.2.2.1 ext		*
			3.2.2.2 int		
				3.2.2.2.1 other trees	*
				3.2.2.2.2 other cc's	*

1: "MEAE" means: "mutually exclusive and exhaustive."

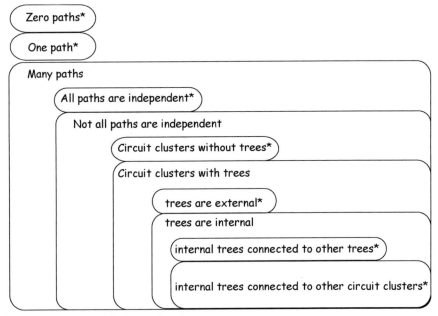

The subsets denoted by * are MEAE

Figure 4-12. Classification chart and Venn diagram for the structures of a bipartite graph. Categories 1, 2, 3.1, 3.2.1, 3.2.2.1, 3.2.2.2.1 and 3.2.2.2.2 are mutually exclusive and exhaustive.

First of all, let us treat the trivial case of trees which terminate with nodes of degree = 1. These are really 1x1 BNSs and do not support the spirit of building multidimensional models. In every case, they represent a relation with a single variable (because the nodal degree = 1) and this variable can be solved for *prior* to its incorporation into the total multidimensional model.

In other words, these "nodal twigs" are single dimensional models and serve only to clutter the total model. They can be easily identified since they are the rows of the constraint matrix which contain only one element. Once that single relevant variable has been determined, its value can be "absorbed" into all its other relevant nodes as a constant, thus eliminated from the constraint matrix and bipartite graph.

Next, we will prove a simple theorem about integers which will help speed many of the proofs in this section:

Theorem 16: Let $p_1, p_2, p_3 .... p_i$ be a set of integers. If $\sum_1^n p_i > -n$ then there exists at least one p_i such that $p_i > -1$.

Proof: Consider the trivial case of $n=1$: obviously if $p_1 > -1$, then $p_1 > -1$. Next, consider $n=2$, then as can be seen in the diagram on the right, if $p_1 + p_2 > -2$, then either $p_1 > -1$, or $p_2 > -1$. Assume that the theorem is true for $n=1$, and we add a $(n+1)$th term equal to -1 (the most stress-full case) to both sides: thus we obtain: $p_1 + p_2 + ... p_n -1 > -n-1$, which is

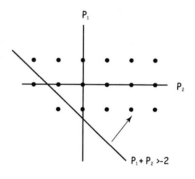

identical to $\sum_1^n p_i > -n$. Thus we have shown that the theorem is true for $n=1$ and $n=2$, and furthermore, if it is true for n, then it is also true for $n+1$. *QED*.

"I believe that this theorem is the least interesting one I have ever seen," sulked the manager.

"I would tend to agree with you," said the analyst. "That's why I frequently refer to it as the 'trivial' theorem and it's not counted among the 'Ten Easy Pieces.' However, as you will soon see, it helps prove other theorems more compactly which will enable us to find BNSs trillions of times faster than the method you suggested earlier (the 'merely' method).

"*Trillions?!*" mumbled the manager under his breath.

We are now prepared to relate the properties of a general BNS to its constraint potential and topological graph structures defined in the previous section by way of these ten compact results:

Theorem 17: If $p(G) = N - K \geq 0$, G contains at least one BNS.

Proof: If G has $p(G) = 0$, its constraint matrix will be a nodal square (NS). If G has $N - K > 0$, its constraint matrix will be a rectangle with more rows than columns, in which case there will always be rows (sub-graphs) which can be removed to bring $p(G) = 0$, thus forming a NS. If there is no smaller NS within it, it is a BNS. If there is a smaller NS within it, it is still a BNS within G. *QED.*

Theorem 18: A BNS cannot have a subgraph, G_s, with $p(G_s)>0$.
Proof: If $p(G_s)>0$, it must have a BNS within it, and by Definition 25 cannot be a BNS itself. *QED.*

Theorem 19: Every BNS must be connected; i.e., it cannot straddle two disjoint connected components.

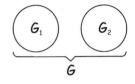

Proof: Assume G is a BNS which is *not* connected and has two subgraphs G_1 and G_2. Since $p(G)=0$, then $p(G_1)+p(G_2)=0$. But then, by Theorem 16, either $p(G_1)>0$, or $p(G_2)>0$. By Theorem 17, G has a BNS within it, contradicting the assumption. *QED.*

Theorem 20: No BNS can be a tree.
Proof: Recalling that terminal nodes are not allowed, the simplest tree is a node between two knots. Adding more tree fragments with N-K<0 will either keep p(G)=-1 or further decrease p(G). Thus p(any tree)<0 and it can't be a BNS. *QED.*

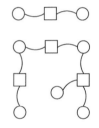

Theorem 21: No BNS can have a subgraph which is a tree.
Proof: Assume that G is a BNS comprised of a tree subgraph, G_t, and another subgraph, G_2, joined at a knot. By this assumption, $p(G)=0$, and from theorem 20, $p(G_t)<0$. Since G_t and G_2 share a knot, $p(G)=p(G_t)+p(G_2)+1=0$, resulting in $p(G_2)>-1$. By Theorem 18, G_2 must have a BNS within in it resulting in the conclusion that G cannot be a BNS since there is a smaller BNS within it. The proof is similar if the subgraphs are connected by a node. *QED.*

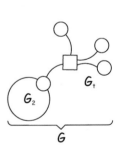

Theorem 22: No BNS can lie across circuit clusters with a separating vertex.
Proof: Assume G is a BNS which lies across circuit clusters G_1 and G_2 with either a node or a knot as a separating vertex. If the vertex is a knot, then $p(G_1)+p(G_2)=-1$; if it's a node, $p(G_1)+p(G_2)=+1$. In either case, by Theorem 16, one or the other of the subgraphs has constraint potential equal to or greater than zero.

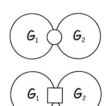

Thus, at least one BNS lies within the subgraphs and no BNS can lie across both. *QED.*

Theorem 23: No BNS can lie across a tree-like network of circuit clusters linked by trees which are attached to trees.

Proof: Assume G is a BNS which lies across a chain of circuit clusters linked by trees. Let G_1 and G_3 be the circuit clusters and G_2 be the tree. Since the total constraint potential equals zero, and the constraint potential of the tree subgraph is negative, by Theorem 16, the constraint potential must be equal or greater than zero in at least one of the circuit clusters. Thus, a BNS must lie in a sub-graph, and the total graph cannot be a BNS. *QED.*

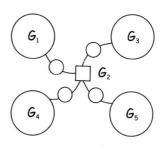

Theorem 24: No BNS can lie across a tree-like network of circuit clusters which are linked by trees to other circuit clusters.

Proof: Since, by definition, the trees linking the circuit clusters together are internal trees, they can have no external twigs and the constraint potential of each tree equals +1, at most. A "meta-tree" can be formed by letting the circuit clusters be vertices and the connecting trees become edges. By Theorem 13, V-E=1 for a tree, thus in the metatree, the number of circuit clusters minus the number of connecting trees equals 1. As before, let us assume that the network is a BNS, so: $p(G) = 0 = \sum_1^n p_i + (n-1)$. This

yields $\sum^n p_i = -n+1 > -n$. Thus, by Theorem 16, at least one of the $p_i > -1$

and thus there must be a BNS within one of the circuit clusters. *QED.*

Theorem 25: Every BNS is the union of adjacent circuits within a circuit cluster. (The "inside out" BNS location theorem.)

Proof: Recall the taxonomy of Definition 35 and Theorem 15 which listed an exhaustive and mutually exclusive set of graph structures. BNSs cannot lie across connected components, can't be a tree or have tree appendages, or be linked by trees, and can't be within circuit clusters linked by separating vertices. The only remaining structure is the union of circuits within a single circuit cluster. *QED.*

Refer to Figure 4-13 for examples of BNSs in graph structures. Note that, although every BNS must lie across adjacent circuits, not every circuit or union of circuits necessarily contains a BNS. This is another example of the non-symmetry of certain theorems which are not "if and only if" theorems. This misunderstanding has been the largest cause of confusion among students.

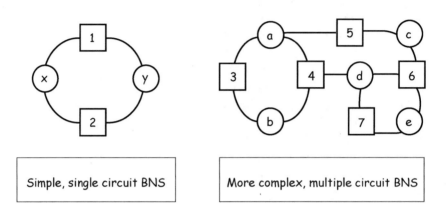

On the right, submodel {34567} is a BNS, and is the union of the circuits: 3a4b, 5c6d4a and 6e7d. These circuits form a circuit cluster since there are at least two independent paths between every pair of vertices. (Note that it would still be a BNS even if there were edges between 3 and d, and 5 and e.)

Note that if **any** of the node were to be removed. the constraint potential of the remaining cluster would be negative and therefore would not be a BNS.

Figure 4-13. Example of Theorem 25, the "inside out" theorem. All BNSs are the union of adjacent circuits and lie completely within a circuit cluster.

Definition 36: A set of BNSs is *independent* if no single BNS is a linear function of any combination of the other BNSs. (See Figure 4-14 for examples.)

Theorem 26: Every circuit cluster (cc) with a constraint potential of $p(cc) > 0$ contains at least $p(cc) + 1$ independent BNSs. (Proof provided later.) Refer to Figure 4-14 for an example of BNSs in cc's.

"Are you done with machine-gunning theorems at me?" complained the manager. "I told you before that even in my most scholarly days, I tended to skip over the proofs of theorems. Here, you haven't avoided *any* proofs and you're even lapsing into using acronyms more frequently. That may speed the exposition, but it also impedes the understanding somewhat."

"Sorry," apologized the analyst, "but I'm really trying to communicate the compactness and rigor of this little corner of mathematics. This hopefully provides a foundation for the rules of the next section. If this were a more typical book on mathematics, the ten theorems would have been aggregated into two or three at the most and the rest demoted to lemmas in longer proofs that would have been far more difficult to follow. Regarding acronyms, it is really worth your effort to learn and master them; it's like using language at a higher level of abstraction. So from now on, we'll use the shorthand 'T12' for 'Theorem 12' and 'D9' for 'Definition 9', etc."

Table 4-1 provides an overview of the ten theorems, using this notation.

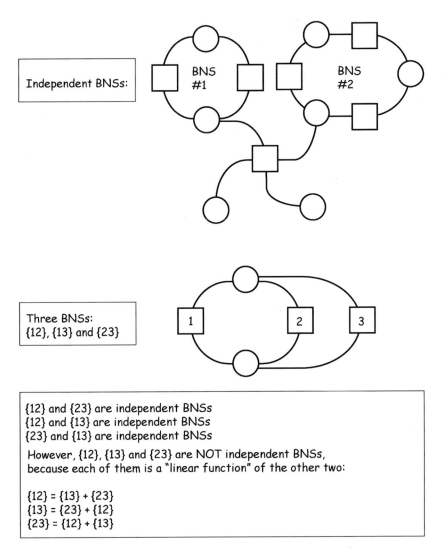

Figure 4-14. Examples of Definition 36. Sets of independent BNSs can have no members which are linear combinations[2] of other members.

---

[2] From the standpoint of their linear algebra representations.

Table 4-1. Ten Easy Pieces Summary; Cornering the BNS within the Bipartite Graph

TABLE 4.1 Ten Easy Pieces Summary; Cornering the BNS within the Bipartite Graph

The constraint potential p(G) is an excellent first-order indicator:

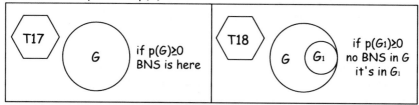

No BNS can exist in any of the following structures:

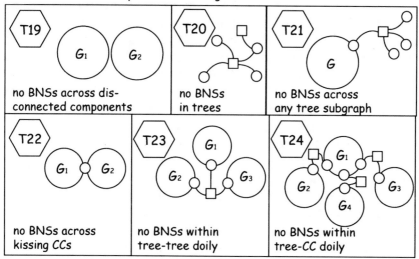

Here is the ONLY structure where BNSs can be found:

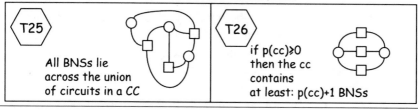

4.5   CONTINUING THE PURSUIT INSIDE THE CIRCUIT CLUSTERS (cc)

The analyst continued: "Now that we have cornered the culprit BNSs to be inside circuit clusters, we must resort to both topological and constraint properties of the bipartite graph to locate them precisely."

"I presume that the constraint potential will be much more valuable here," the manager ventured.

"True," agreed the analyst, "but not nearly as valuable as we would like. For example, one would hope that if $p(cc)<0$, we could be assured that there are no BNSs within that cc, and if $p(cc)=0$, then there are exactly $p(cc)+1$ BNSs within that cc. Unfortunately, we cannot reach either of these two conclusions. Referring to Figure 4-15a, we see that a cc with $p(cc)<0$ can still have BNSs and a cc with $p(cc)>0$ can have any number of BNSs equal or greater than $p(cc)+1$. Look at the outer ring of the cc; because it is a bipartite graph, $p(\text{outer ring})=0$. We can form Y inner loops, each of which increases $p(cc)$ by 1, while at the same time we can form Z other inner loops each of which decreases $p(cc)$ by 1. Thus, Y additional BNSs are formed but the $p(cc) = Y-Z$. Rearranging, we see that $Y=p(cc)+Z$ and that the number of BNSs$=Y+1=p(cc)+Z+1$. Since $Z=0$, we conclude: the number of BNSs$=p(cc)+1$. (This is the proof of T26 promised above.)"

"No more use than that?" complained the manager.

"Well, we can squeeze out a little more: From the viewpoint of the constraint matrix, if a cc has $p(cc)>0$ then it has N-K more rows than columns and thus N!/(N-K)!K! nodal squares can be found within this rectangle (Figure 4.15b). That's a help, but beware; just because every NS must contain at least one BNS does not allow us to conclude that there are at least N!/(N-K)!K! BNSs in the cc. In Figure 4.15c, we see that a single BNS may be common to two or more NSs, and in Figure 4.15d, we see there can be more than one BNS in an NS."

"So we must also resort to the topology of the cc. From T25, we know that every BNS lies across the union of circuits and from T14, we know that the number of independent circuits in a cc is its circuit rank, $c(cc)$. This allows us to establish the maximum number of BNSs in a cc as the power set of the cc's circuit rank:

Theorem 27: The maximum number of BNSs in a cc equals the power set of the circuit rank of the cc: Max # of BNSs $= 2^{c(cc)}$

"Thus, T26 sets the lower limit on the number of BNSs as a function of the cc's constraint potential and T27 sets the upper limit as a function of the cc's circuit rank," the analyst concluded.

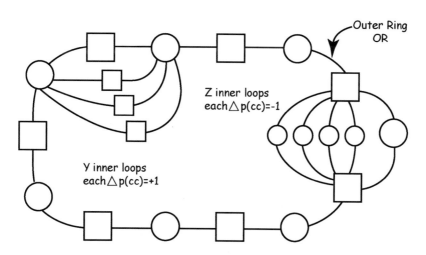

Start with the outer ring (OR) of a circuit cluster which has p(OR)=0.
By definition, this is a BNS.

Y inner loops can be formed, each one of which is attached only to
knots, and each of which **increases** p(cc) by +1 and creates a new BNS.

Z inner loops can also be formed, each one of which is attached only to
nodes and each of which **decreases** the p(cc) by 1. None of these will
create new BNSs.

Thus, Y additional BNSs are formed and p(cc)=Y-Z; thus Y=p(cc)+Z
The number of BNSs = Y+1 = p(cc)+Z+1
Since Z=0, then the number of BNSs = p(cc)+1

Figure 4-15a. Demonstration of T26: the number of BNSs in a circuit cluster is equal or
greater than p(cc)+1.

"So what do you suggest a practical guy to do?" queried the manager.

"Constraint theory provides at least two complementary robust
procedures for locating the BNSs within the cc's:

a)   The *brute force procedure* which merely examines every submodel
     within the cc and checks for p(submodel)=0, the definition of a
     BNS.
b)   The *T25 procedure* employing the fact that every BNS must be the
     union of adjacent circuits.

b) The constraint matrix of a circuit cluster, cc, with p(cc)>0

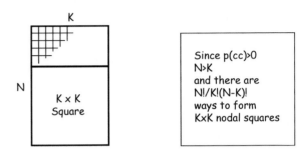

Since p(cc)>0
N>K
and there are
N!/K!(N-K)!
ways to form
KxK nodal squares

c) The BNS {1.2} is shared by the NSs {1,2,3,4} and {1,2,5,6}

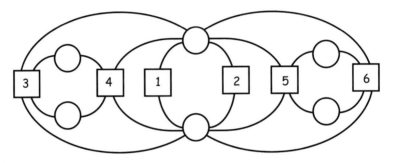

d) The NS {1.2,3,4,5,6} contains two BNSs: {1,2} and {3,4}

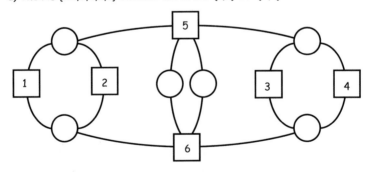

Figure 4-15 b-d. In a circuit cluster with p(cc)>0, it is easy to determine the number of nodal squares within it, but this does not lead directly to the number of basic nodal squares.

The *brute force procedure* develops the power set of the N relations – each one being a submodel – and tests for p(submodel)=0 and that there are no submodels with p(submodel)=0 within it. There will be 2^N such submodels to test and the testing of each one is simple in the extreme. True, we seem to be trapped again in an exponential number of trials but in general the number of nodes in even the largest of circuit clusters will be far smaller than the number of nodes in the total model. For example, if the total model had 100 relations, it would be very unusual to have a circuit cluster as large as 30 relations. The computational advantage of performing brute force testing within a cc rather than within the total model equals $2^{100}/2^{30} \sim 10^{21}$ – truly enormous! Assuming that each test can be accomplished in a nanosecond, the brute force examination of each of the cc's submodels would require only a second.

The *T25 procedure* offers greater efficiency as the number of relations within the cc grows much larger than 30. By definition, it makes use of the fact that every BNS is the union of adjacent circuits. The first step in this procedure is to develop the set of independent circuits within the cc, which by T14 is equal to the cc's circuit rank, c(cc). Then examine the submodels which reside on these circuits – including the twig knots which may be attached – one at time, then two at a time with adjacent circuits, then three at a time with adjacent circuits, etc. The *disadvantage* of this procedure is that we must find the independent set of circuits and form the potential sets of combinations of these circuits, before we test for zero constraint potential of any of the submodels. The *advantage* of this procedure is that, instead of searching through the power set of the N relations of the cc, we are searching through the power set of the cc's independent circuits. Assuming again that N=30 and that c(cc)=6 (an average of 5 relations per circuit), then the computational leverage is about $2^{30}/2^{6}$, a factor of several million. Of course this must be weighed against the additional "overhead" of forming the independent set of circuits and its power set.

Generally speaking, it is judged the brute force procedure would be more practical for cc's with N<30 if it could be accomplished on the order of a second. As N grows to 50 and beyond for the cc, it would become increasingly attractive to employ the T25 procedure. (Refer to Figure 4.16).

4.6   LOCATING BNSs WITHIN A MODEL GRAPH

Let us now apply the definitions and theorems of the previous sections to the challenge of finding BNSs within a model graph.

> **_Definition 37:_** General procedure for locating BNSs in a model graph, which involves the sequential application of these "sieves": connected components, tree structures, circuits, circuit clusters and constraint potential.

This procedure is outlined in the following steps a) – h) and illustrated graphically, with a notional model graph, in Figures 4-16 through 4-21.

Step a) trims down the model by eliminating all nodes of degree one. These twig nodes represent single-variable equations which can be solved separately by employing the full model. Without this key topological property, the cornerstone theorems ("ten easy pieces") of constraint theory could not have been proven. As a subset of the "ten easy pieces", T-21 through T-25 are specifically employed in the D-37 process. As such, no terminal nodes are allowed before the rules of constraint theory can be applied. Not only are the terminal nodes trimmed from the model graph but their incident edges, which would become "dangling", must also be cleaned up explicitly.

Once the relevant single variables (knots) of the above terminal nodes have been determined, analytically or numerically, their values can be incorporated into other relevant relations as coefficients. In effect, these are 1×1 BNS which only serve to clutter a multi-dimensional model and can be temporarily eliminated. The solution time for this process is linear as each and every node will be examined once, and only once, for checking of $d(n) = 1$ and removal as such. It should be clarified that the terminal nodes are only temporarily eliminated, or hidden, for the purpose of topological simplification while we are searching for other higher-order BNS in the subsequent steps of D-37. Afterwards, these terminal nodes will still need be re-integrated back into the original, complete model to check for possible over-constraint among all BNS (Phan, pp. 101-104).

Step b) decomposes the total graph into connected components by employing graph theoretic concepts of spanning tree and full-spanning forest to improve solution time over that of D-27. In particular, to identify each component in a model graph, the widely-accepted depth-first search (DFS) algorithm is used to grow a spanning tree. This algorithm was originally devised by Tarjan (1974), [27] and has been extensively referenced by Cormen et al. (2001), Dechter (2003), Gross & Yellen (2006) and many others in the literature. The DFS also lends itself naturally as an effective and efficient application in steps (d) and (e) of D-37 and thus can even save more solution time (Phan, pp. 104-114).

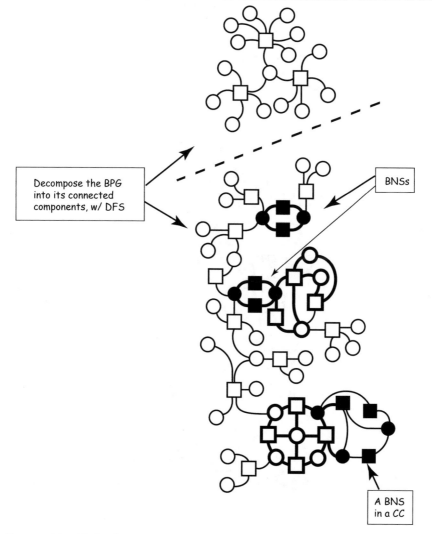

Figure 4-16. DFS algorithm used to decompose the BPG into two connected components.

Step c) discards those components without any circuit as identified in Step (b) since, per T-25, purely tree-like structures cannot contain any BNS. D-31 stipulates computation of the circuit rank, c(G), for every component and discard those with c(G) = 0. However, the inherent products coming out of Step (b) include only those components with c(G) > 0. Thus, repeated calculation of c(G) is not necessary (Phan, pp. 114-115).

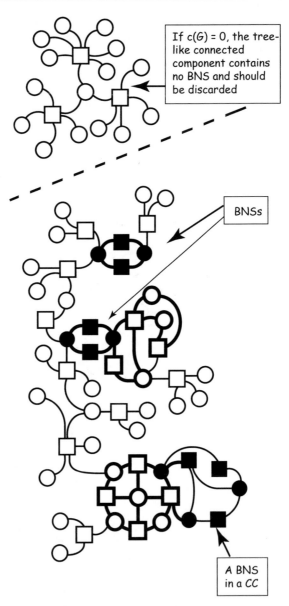

If c(G) = 0, the tree-like connected component contains no BNS and should be discarded

BNSs

A BNS in a CC

Figure 4-17. Retain only the lower connected component with c(**G**) > 0.

Step d) uses D-30 to trim repeatedly each spanning tree of every external twigs, i.e. vertex with $dC(v) = 1$, since no BNS can exist in tree twigs by T21. The symbol $dC(v)$ is meant to emphasize the degree of a vertex within the context of its parent component coming out of Step (c), and not that of the spanning tree which is just a subgraph of the component. Even though this step only operates on spanning trees, it is important to keep in mind that the relative complement of each spanning tree must always be carried and fully accounted for, as an imperative part of every component, to be simplified and decomposed through every sequentially-related step within the D-37 process (Phan, pp. 115-117).

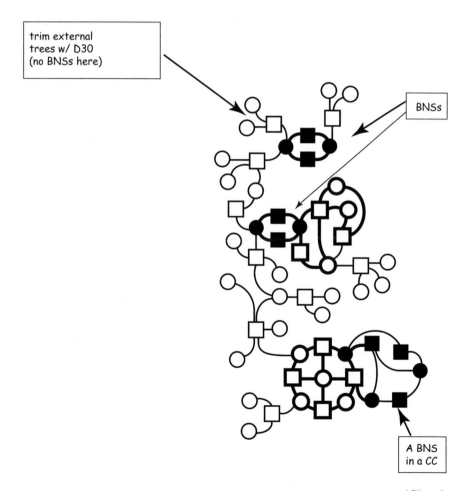

Figure 4-18. D-30 used to trim external trees from a connected component with $c(\mathbf{G}) > 0$.

Step e) can employ one of several methods to identify and remove all internal trees from every spanning tree of each component coming out of Step (d) since no BNS can exist within internal trees by T23 and T24. A spanning tree will be decomposed into smaller spanning trees, corresponding to newly-formed respective sub-components. The relative complement of each resultant smaller spanning tree will also be automatically identified as by-products. In an edge-centric method, a spanning tree needs to be grown only once for a component to save solution time (Phan, pp. 117-122). Other methods to determine internal trees include those by Tarjan, Cormen et al. (pp.558-559) and Gross & Yellen (pp. 182-184).

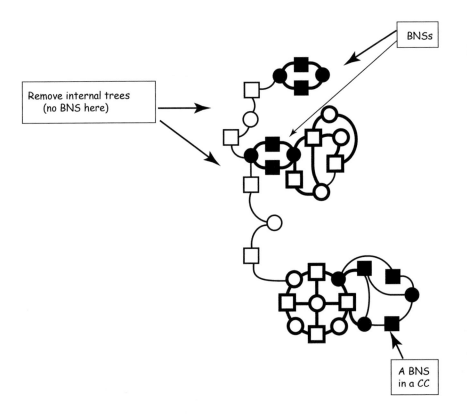

Figure 4-19. All internal trees to be removed from a connected component with c($\mathbf{G}$) > 0.

Step f) identifies separating vertices of every BPG coming out of Step (e) and partitions its kissing circuits, or circuit clusters, at such points. D-28 stipulates a trial-elimination procedure with polynomial runtime of $\Theta(V \cdot E)$, where V and E are the numbers of vertices and edges, respectively (Shirey, 1969). However, a more efficient DFS-based algorithm with linear runtime can be used to separate kissing circuits and circuit clusters. The notion of a separating vertex in Constraint Theory (D-26) is identical to that of a cut-vertex, cut-point or articulation point in Graph Theory. And the structure of a circuit cluster in Constraint Theory is the same as that of a biconnected component in Graph Theory (Phan, pp. 123-143). For each partitioned circuit cluster, if p(cc)>0 then the cc has at least p(cc)+1 BNS within it.

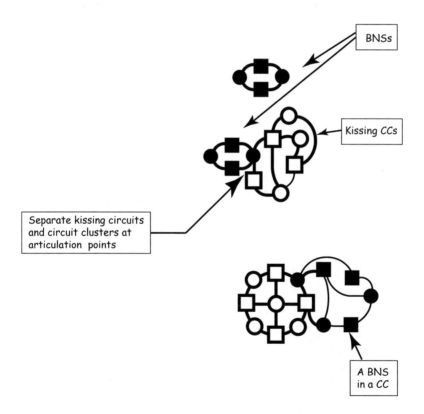

Figure 4-20. Kissing circuits and circuit clusters to be separated at articulation points.

Step g) can follow one of several methods to search for intrinsic BNS within each cc coming out of Step (e). For smaller cc with $N \leq 14$, the brute-force approach is to examine its nodal power set, which would result in an exponential solution-time of 2^N (Phan, 2011, p. 381). For larger cc, a more efficient method for locating potential BNS is to examine only the fruitful unions of overlapping nodes, either directly or transitively, i.e. avoiding combinations of non-overlapping nodes. The search for BNS should also be systematically implemented in a bottom-up approach, smaller nodal unions before larger ones. As such, smaller nodal unions identified as BNS, or BNS containers, can be tagged to not be re-used as components in the construction of any larger union. Such a larger union can never be a BNS and should not be unnecessarily constructed and examined (Phan, pp. 144-231).

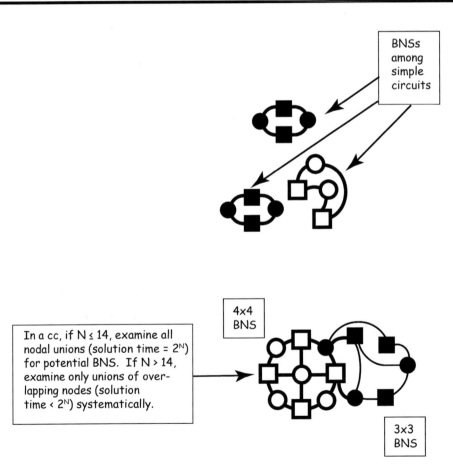

Figure 4-21. Search for BNS among simple circuits and within circuit clusters.

> **Step h)** validates the potential BNSs located by step (g) in simplified circuit clusters, which are isolated sub-graphs of the overall original model graph. These potential BNS must still be validated, as complete sub-models, within the context of a connected component of the model graph identified in step (b). The validation process is implemented by re-integration of tree twigs pruned in step (d), internal trees removed in step (e) and circuit clusters decomposed at separating vertices in step (f) (Phan, pp. 323-325).

"Surely, for those not familiar with D37, there must be other ways of managing high dimension models on computers", insisted the manager.

"Yes, there are ways," admitted the analyst. "But they require far more attention by the people developing the model and programming it than one would like in this world of highly computer-automated analysis that we like to believe we're in. A pragmatic approach would be to start with models whose constraint potential is enormously negative – which have perhaps a 100 or so fewer equations than variables. Then, once the computational request is made known to the model manager, the potential computational paths – which started out under-constrained – are examined to see where variables can be set at constant values so that the constraint potential is brought up to zero along the entire path. This would be very labor intensive and fraught with the danger that the programmers would make these 'constraint enhancing' decisions without properly consulting those with the responsibility of model fidelity. I have a very sad story in my past – among many others I'm certain – wherein a programmer decided to add a key constraint to the model without informing management, and as a result my company lost a crucial contract we had worked on for years. Typically, the math model results were presented to management with a minimum of visibility."

"Getting off the war stories for a moment," said the manager changing the subject, "doesn't step f) of D37 leave us hanging a bit? It says that if $p(cc)>1$ then the cc has multiple BNSs. What happens then?"

"Thanks for the question," the analyst said. "We have spent most of this chapter on the issue of the location of BNSs but, as crucial as that is, it is still only part of the job of math model management. It's time to broaden our sights to the whole problem."

4.7 QUERIES FOR THE REGULAR STUDENT

1. Consider the following mathematical model of regular relations:
   $a = f_1(c, d)$, $h = f_2(g, m, w)$, $d = f_3(r, g)$, $c = f_4(a)$, $d = f_5(c)$,
   $t = f_6(r, s)$

Is the model connected?
Is it tree-like, circuit-like or both?
What is the circuit rank of the model?
What is the constraint potential of the model?
Are there BNS's in the model? If so, identify it (or them).

2. Consider the following mathematical model of regular relations:
    $x = f_1(y), z = f_2(w, y), y = f_3(x), x = f_4(y)$
    What is the circuit rank of the model?
    What is the model's constraint potential?
    Are there circuit cluster(s) in the model? If so, identify it (or them).
    Are there BNS's in the model? If so, identify it (or them).

3. Consider the following math model of regular relations:
    $m = f_1(r, p), n = f_2(q, r), r = f_3(m, n), q = f_4(p, r)$
    Is there a circuit cluster in this model? If so, identify it.
    What is the model's circuit rank?
    What is the total number of simple circuits in the model?
    What is the number of independent circuits?
    What is the model's constraint potential?

4. Provide a shorter proof than that given in the text for Theorem 21, using only the constraint matrix.

5. If "sc" refers to a simple circuit, prove that $p(sc) = 0$. If "nsc" refers to a non-simple circuit, show examples how $p(nsc) > 0$ and $p(nsc) < 0$.

6. Prove that, in a circuit cluster, the number of independent circuits is equal to or greater than the number of BNSs.

7. Draw the bipartite graph of a circuit cluster which is a nodal square that contains two BNSs within it.

8. Draw the bipartite graph of two overlapping nodal squares which share a single BNS within them.

9. There is an easy way to count independent circuits much of the time: merely locate all the "white areas" inside the bipartite graph which are completely surrounded by edges. Why will this method not work in general?

Chapter 5 MODEL CONSISTENCY AND COMPUTATIONAL ALLOWABILITY

5.1 ZERO CONSTRAINT ALL ALONG THE COMPUTATIONAL PATH

Now that we have developed a general process to locate intrinsic BNSs, we can return to the concepts expressed in Figure 4-7 and discuss the general issue of constraint propagation through a connected graph of regular relations. The general rule, as depicted in Figure 5-1, is compactly stated as:

In order for a computational request on a consistent model to be allowable, the entire computational path, from independent variables and constants to dependent variable, must have a resultant constraint potential of zero.

If the resultant constraint potential exceeds zero at any point, the computation is *over-constrained*; if the resultant constraint potential is less than zero at any point it is under-constrained. It is possible for the same computational request to be both over- and under-constrained – at different places along the computational path. In short, the resultant constraint potential must be *just right* along the entire path. Thus, the designation: *The Goldilocks rule.*

Referring to Figure 5-1, examine the nodes first. Recall that the local degree of any node, $d(N)$ is simply the number of edges that intersect that node. The intrinsic constraint potential of that node $p_i(N)$ is by definition $N - K = 1 - d(N)$. Now if constraint flows *into* this node from elsewhere in the model, $p_i(N)$ will be increased by $I(N)$, the number of edges which propagate constraint into the node. The resultant constraint potential then becomes:

© Springer International Publishing AG 2017
G.J. Friedman, P. Phan, *Constraint Theory*, IFSR International Series on Systems Science and Engineering, DOI 10.1007/978-3-319-54792-3_5

$$p_r(N) = p_i(N) + I(N) = [1 - d(N)] + I(N)$$

Let $I(N) = d(N) - 1$; we see that this drives $p_r(N)$ to zero, and we have just derived the "$(d - 1)$ in / 1 out" rule for nodes.

Looking at:	If:	Then:	ref below
Nodes	$P_r(N){>}0, I(N){>}d{-}1$ $P_r(N){<}0, I(N){<}d{-}1$ $P_r(N){=}0, I(N){=}d{-}1$	overconstraint underconstraint perfect constraint	a b c
Knots	$P_r(K){>}0, I(K){>}1$ $P_r(K){<}0, I(K){<}1$ $P_r(K){=}0, I(K){=}1$	overconstraint underconstraint perfect constraint	d e f
Circuits	$P_r(circuit){>}0$ $P_r(circuit){<}0$ $P_r(circuit){=}0$	overlapping BNSs no BNSs one BNS	g h i

Legend: P_r(vertex) is the resultant constraint potential at that vertex
I(vertex) is the number of constraint inputs to that vertex

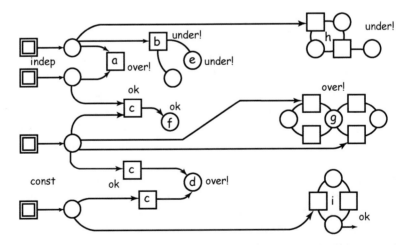

Figure 5-1. The Goldilocks rules for constraint flow through a network. In order for a computational request to be allowable, the resultant constraint must equal zero at every vertex and circuit along the computational paths from independent variables and variables held constant to the dependent variable.

Next, examine the knots of Figure 5-1. The intrinsic constraint potential of a knot is by definition: $P_i(K) = N - K = d(K) - 1$. If this knot produces $O(K)$ outputs, then the resultant constraint potential of the knot is decreased by $O(K)$: $p_r(K) = p_i(K) - O(K)$. Recognizing that $O(K) + I(K) = d(K)$, we see that setting $O(K) = d(K) - 1$ will drive $p_r(K)$ to zero. Thus we have just derived the "1 in / (d – 1) out" rule for knots.

Finally, examine the circuits on the Figure 5-1 computational paths. As we have seen above, if p(circuits) > 0, we will have multiple BNS and therefore over-constraint on their common variables. On the other hand, if p(circuits) < 0, then we will experience under-constraint. If p(circuits) = 0, then a single BNS will provide constraint at all its variables, permitting the flow of constraint though these circuits.

In summary, the "Goldilocks" rule stating the necessity of zero constraint all along the computational path is a unifying concept for computational allowability.

5.2 RECAPITULATION OF COMPUTATIONAL FLOW

Let us attempt a recapitulation of what is involved in a computational request at this point.

The first order of business is to determine the model's consistency, for if it is inconsistent then no computational request will be allowable. All disconnected components and tree structures of universal relations are intrinsically consistent. However, in circuit-like bipartite graphs there may exist one or more BNSs which will exert intrinsic constraint on all of their relevant knots. A single BNS merely restricts the number of computational requests, but multiple BNSs often drive the model into inconsistency and thus prevents all computational requests. These over-constraints must be relieved before any computational requests can be entertained. Furthermore the constraint flowing out of each BNS may intersect in a larger constraint flow domain, and will also require relief by the model builders.

Once the model's consistency has been established, computational requests on it may be examined. The general format of these requests will be: "Please compute the dependent variable X as a function of the set of independent variables {Y}, with the set of {Z} variables held at the constant values {Z_o}."

As part of the consistency check, the bipartite graph already has its domains of intrinsic constraint mapped out. For each computational request, add the *extrinsic* constraint sources due to the independent variables {Y} and the variables held constant {Z}. The effect of applying extrinsic constraint to the variables held constant {Z} will be to cause them to disappear from

Constraint theory

the graph completely (the $\{Z_o\}$ constants will be "absorbed" into their relevant nodes as parameters in their equations rather than variables). This action will tend to simplify the model and sometimes even disconnect it, making computations across disconnected components impossible. See Figure 5-2.

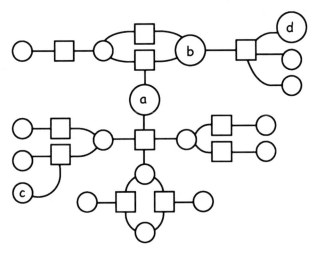

(a) BPG prior to the application of holding variables constant

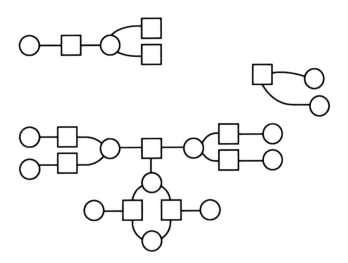

(b) BPG after the variables a, b, c and d are held constant

Figure 5-2. Applying extrinsic constraint by holding some variables constant tends to simplify the bipartite graph. Sometimes the graph is disconnected into separate components, rendering computational requests involving variables in different components unallowable.

Next, start the flow of extrinsic constraint from each of the {Y} independent variables into the remainder of the bipartite graph. Several circumstances may occur, as is shown in Figure 5-3.

a)   In the vicinity of tree structures, the (d − 1) in / 1 out rule for nodes and the 1 in / (d − 1) out rule for knots will always be sufficient to determine over or under constraint.

b)   Even in the vicinity of some circuits, these rules will still be sufficient.

c)   On many occasions, the constraint will flow into a circuit structure and appear to stop. But closer examination shows that the constraint flow into a node effectively increases the resultant constraint potential of the circuits, forming them into a resultant BNS, which by applying constraint to all its relevant knots, permits the constraint flow to continue.

d)   Sometimes, the intrinsic and resultant BNSs reside in close proximity.

See Figure 5-4 for examples of interactions in these cases.

5.3   GENERAL PROCEDURE FOR DETERMINING CONSISTENCY AND ALLOWABILITY IN A MODEL OF REGULAR RELATIONS

"Recall that in Chapter 1, the existence of a single BNS was an irritant; you were disappointed that you couldn't accomplish all the computational requests you desired, but it didn't drive the model into inconsistency," the analyst continued. However, the existence of overlapping BNSs will cause the model to be intrinsically inconsistent; therefore no computations at all are allowed. That is not just irritating, it's devastating. Multiple, overlapping BNSs are against nature."

"My, my; in addition to being a dealer in hyperbole," commented the manager, "you're waxing philosophical, too. What exactly do you mean by 'against nature'?"

"I mean at least two related things," responded the analyst, "which can be captured in the following postulate:"

Postulate 2: First, the laws of physics and other descriptions of the world are fundamentally consistent if they are fully understood, and second, it is the intent of model builders to represent phenomena accurately and thus if over-constraint occurs, it is unintentional.

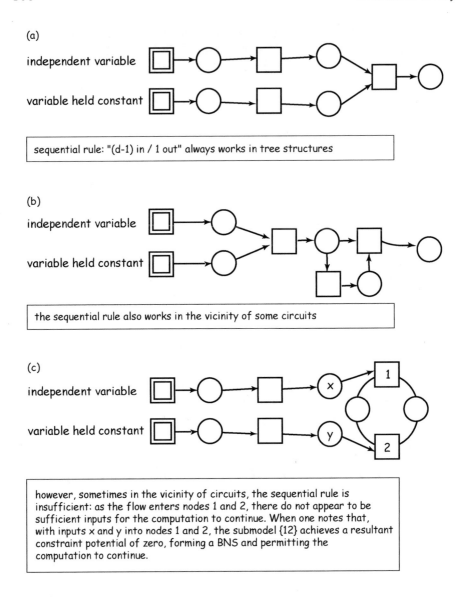

Figure 5-3. Flow of constraint in the vicinity of trees and circuits.

"So, I was referring to the nature of the world as well as the nature of model builders. Since the natural world itself is fundamentally consistent, over-constraint is invariably contributed by the fragmented understanding of the human model builders, either by inadvertently applying excessive relations to the description of a phenomenon or, more frequently, adding too many relations of policy, design rules, optimization criteria or desired outcome."

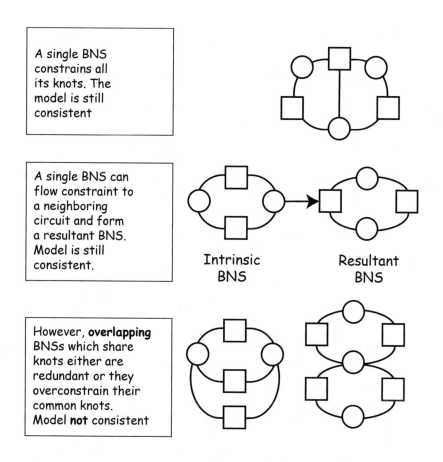

Figure 5-4. Examples of constraint flow in the vicinity of BNSs.

In order to bring a circuit cluster with p(cc) to consistency, at least p(cc) nodes must be removed. In other words, of the N nodes in the cc, N − K of them must be removed. From simple combination and permutation theory, we conclude that there are N!/(N−K)!K! ways to remove N − -K nodes from the original N nodes. Unfortunately, these removals lie outside the domain of Constraint Theory and require the group of human model builders who contributed to the circuit cluster. Negotiations of the type discussed in Chapter 1 will be required and the discussions could get tense. Better that than having no working model at all!

We are now prepared to describe the procedure of determining consistency and computational allowability in a bipartite graph of regular relations:

Definition 38: General procedure for regular relations.

a)   Locate the BNS by employing **_D-37_**. If no BNS found, including 1 × 1 BNS (a twig node and its relevant knot) and c(**G**) = 0, the model is inherently consistent. Skip steps (b) – (f), go to step (g).

b)   If overlapping BNSs are found, eliminate nodes through negotiation so that all remaining BNSs are non-overlapping.

c)   Propagate constraint emanating from the BNSs to their resultant constraint domains, employing T10: for nodes, d(n)-1 in and 1 out, and for knots, 1 in and d(k)-1 out.

d)   If any two of these resultant constraint domains overlap, determine whether any knots are over-constrained. If so, by negotiation, remove sufficient nodes to relieve the over-constraint.

e)   After the resultant domains have expanded as far as T10 applies, analyze the remaining graph for resultant BNSs, using D36. If new BNSs arise, continue with T10 and then apply D36 again. Continue until the constraint domains no longer increase.

f)   If all the overlapping BNSs are reconciled and resultant constraint domains expand without over-constraint, then the model is CONSISTENT.

g)   For each computational request, treat all independent variables and variables held constant as extrinsic sources of constraint which are added to the intrinsic sources of constraint developed above.

h)   Propagate computational paths from all constraint sources throughout the model employing the T-10 rules. If any knot or node is over-constrained, the computational request is NOT ALLOWABLE.

i)   If the computational path does not reach the dependent variable, then examine the residue for BNSs, using D36, and continue with T10, followed by D36 as necessary until the computational path can go no further.

j)   If the path does not reach the dependent variable, or over-constrains a knot or node, the computational request is NOT ALLOWABLE.

k)   If the path reaches the dependent variable by employing all independent variables, without either over- or under-constraint along the way, then the computational request is ALLOWABLE. It is acceptable to have local under-constraint elsewhere in the model. If

> the dependent variable can be reached without having to propagate the constraint externally imposed on one or more independent variables, then the computation is UNALLOWABLE.

Steps (a) – (f) address the issue of model consistency, and steps (g) – (k) that of computational allowability. The following sections of this chapter will examine the above general procedure in more details, step by step, with the goal of realizing the utility of Constraint Theory. D-37 and D-38 will be further extended, refined and improved into a set of more effective and efficient algorithms, ready for implementation. As important as these issues are to discuss, they do not represent exponential explosions of computational time.

5.4   DETECTION OF OVERLAPPING BNS

Step (a) of **D-38** states that: "Locate the BNS by employing **D-37**. If no BNS found, including 1×1 BNS (a twig node and its relevant knot) and $c(G) = 0$, the model is inherently consistent. Skip steps (b) – (f), go to step (g)."

Per **T-25**, all BNS exist within simple circuits or across clusters of adjacent circuits. If a model graph **G** with circuit rank $c(G) = 0$, it has no circuits within. As such, **G** contains only tree structures. By **T-9**, any set of universal relations whose BPG has a tree structure is consistent. As a global criterion, and a quick check, to distinguish between trees and circuit clusters, **T-13** asserts that "*a connected model graph having $V - E = 1$ is a tree*". This assertion comes directly from the definition of circuit rank (Phan, pp. 31-32).

Otherwise, after all the intrinsic BNS have been identified and validated within the context of a connected component in step (h) of **D-37**, they need be compared against one another for any *direct over-lapping*. To check for over-lapping, a set of definitions and vectorial operations will be developed and illustrated herein.

Definition 39: Two BNS are said to **_directly overlap_** if they share at least one variable in common, i.e. the intersection of their knot sets is non-null. In other words, the common variable(s) are said to be **_over-constrained_** (or **_over-specified_**), which causes the parent model graph to be inherently inconsistent.

Figure 5-5 illustrates a model graph **G** and its constraint matrix [C_G]. Within **G**, three intrinsic BNS can be identified:

✓ BNS #1 consisting of nodes 6 and 7, and their relevant knots e and f.
✓ BNS #2 consisting of nodes 1 through 6, and their relevant knots a through f.
✓ BNS #3 consisting of nodes 1 through 5, and 7, and their relevant knots a through f.

Since these intrinsic BNS overlap one another, i.e. sharing and thus over-constraining at least one common variable, **G** is inherently inconsistent and no computational requests made on **G** are allowable.

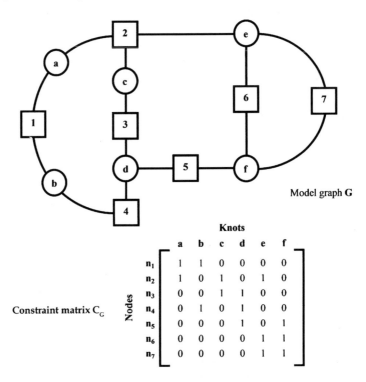

Figure 5-5. Inherently inconsistent model graph with multiple intrinsic BNS overlapping one another.

Definition 40: The characteristic vector (or *charvec*) of a BNS is the union, based on the bitwise *inclusive OR* operation, of all the charvecs of its nodes (see Definition D-9 for *charvec*, and Definition D-10 for bitwise *inclusive OR*). Symbolically,

$$charvec(\text{BNS}) = \bigcup_{i=1}^{|N_{BNS}|} charvec(n_i) \qquad (5-1)$$

In the constraint matrix **[C_G]** of Figure 5-5,

charvec(n_6) = (0, 0, 0, 0, 1, 1)
charvec(n_7) = (0, 0, 0, 0, 1, 1)
$\left.\rule{0pt}{20pt}\right\}$ charvec(BNS$_1$) = charvec(n_6) $\cup$ charvec(n_7) = (0, 0, 0, 0, 1, 1)

charvec(n_1) = (1, 1, 0, 0, 0, 0)

charvec(n_2) = (1, 0, 1, 0, 1, 0)

charvec(n_3) = (0, 0, 1, 1, 0, 0)

charvec(n_4) = (0, 1, 0, 1, 0, 0)   charvec(BNS$_2$) = charvec(n_1) $\cup$ ... $\cup$ charvec(n_6)

charvec(n_5) = (0, 0, 0, 1, 0, 1)   = (1, 1, 1, 1, 1, 1)

charvec(n_6) = (0, 0, 0, 0, 1, 1)

charvec(n_1) = (1, 1, 0, 0, 0, 0)

charvec(n_2) = (1, 0, 1, 0, 1, 0)

charvec(n_3) = (0, 0, 1, 1, 0, 0)
charvec(BNS$_3$) = charvec(n_1) $\cup$ ... $\cup$ charvec(n_5) $\cup$ charvec(n_7)
charvec(n_4) = (0, 1, 0, 1, 0, 0)
                       = (1, 1, 1, 1, 1, 1)
charvec(n_5) = (0, 0, 0, 1, 0, 1)

charvec(n_7) = (0, 0, 0, 0, 1, 1)

Definition 41: The **BNS matrix** of a bipartite graph **G**, denoted Ω_G, is the rectangular array of binary numbers whose rows correspond to the BNS and columns correspond to the knots in **G**. In Ω_G, the entry $e_{i,\,j} = 1$ if knot $k_j \in$ **K_G** has been specified by BNS$_i$, and 0 otherwise.

Figure 5-6 presents the BNS matrix Ω_G for the example BPG in Figure 5-5. Note that each row of Ω_G represents the characteristic vector of the corresponding BNS.

Knots

	a	b	c	d	e	f
BNS₁	0	0	0	0	1	1
BNS₂	1	1	1	1	1	1
BNS₃	1	1	1	1	1	1

Figure 5-6. BNS matrix Ω_G of BPG in Figure 5-5.

Definition 42: The **_overlapping factor_** between BNS_i and BNS_j, denoted as $\omega_{i,\,j}$, is defined as the dot product between their characteristic vectors. Symbolically,

$$\omega i, j = charvec(BNSi) \bullet charvec(BNSj) \qquad (5-2)$$

Furthermore, by definition of dot product, the overlapping factor $\omega_{i,\,i}$ also indicates the number of common variables (knots) over-constrained by the two BNS.

Theorem 28: BNS_i and BNS_j are overlapping if $\omega_{i,\,i} \geq 1$. They are not overlapping if $\omega_{i,\,i} = 0$.

Proof: If two BNS share any common knot in k_i then the corresponding i^{th} elements in both of their respective characteristic vectors are equal to one. Therefore, the dot product between their characteristic vectors must be equal to one, or larger, if there are one, or more, common knot between the BNS. *QED*

In Figure 5-6, $\omega_{1,\,2} = charvec(BNS_1) \bullet charvec(BNS_2) = (0, 0, 0, 0, 1, 1)$ $\bullet$ $(1, 1, 1, 1, 1, 1) = 2$. Thus, BNS_1 and BNS_2 overlap by two variables, which confirms a visual inspection of Figure 5-5. This technique of employing dot product provides a quick check for potential overlapping between two BNS. Once any overlapping has been confirmed, the over-constrained variables can be identified via an element-by-element comparison between the BNS charvecs.

Definition 43: Algorithm to detect overlapping BNS in a model graph and to identify variables commonly shared among them.

      Input: A matrix Ω_G with R rows and K columns representing all BNS in a graph **G**.

Output: A subset $\mathbf{S} \subseteq \mathbf{K_G}$ consisting of overlapped knots (over-constrained variables).

> Initialize $\mathbf{S} = \emptyset$.
> For i ranging from 1 to $(R - 1)$
>> For r ranging from $(i + 1)$ to R
>>> If $\omega_{i, r} \geq 1$
>>>> For j ranging from 1 to $|\mathbf{K_G}|$
>>>>> If the jth element of BNS_i = that of BNS_r = 1
>>>>> Add the corresponding knot k_j to $\mathbf{S}$.
> Return $\mathbf{S}$ (if $\mathbf{S} \neq \emptyset$, the model is inconsistent).

Theorem 29: To detect overlapping BNS and identify over-constrained variables using a matrix Ω_G with R rows and K columns representing all BNS in a model graph $\mathbf{G}$, the required solution time will be:

$$\Theta_{detection} = K\ (R^2 - R/2) \sim K \cdot R^2 \qquad (5 - 3)$$

Proof: The matrix Ω_G has R rows representing all BNS and K columns representing knots (variables) in $\mathbf{G}$. There are $(R - 1) + (R - 2) + \ldots + 2 + 1 = R^2 - R/2$ pairwise comparisons of BNS. For each pairwise comparison of BNS, all K columns must be examined to look for commonly shared, and thus over-constrained, variable(s) between the two corresponding BNS. *QED.*

The above algorithm needs be applied to all BNS, including those 1×1 BNS born of terminal nodes initially trimmed by step (a) of **D37** process defined in Section 4.6. Figure 5-7 illustrates an example model graph back into which a previously trimmed terminal node (nodal twig) is re-integrated as a 1×1 BNS. This re-integration reveals knot c as over-constrained between $BNS_1 = \{ 3, c \}$ and $BNS_2 = \{ 1, c, 2, b \}$.

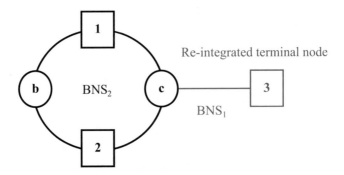

Figure 5-7. Re-integration of previously trimmed terminal node reveals model inconsistency

At this point in the process, the model builder needs to examine the set of over-constrained variables output by the algorithm in **Definition 43**. Any inconsistency, i.e. directly overlapping intrinsic BNS, inherent within the model graph must be resolved <u>before</u> proceeding with the **D-38** process. This requirement for reconciliation has been established in steps (b) and (d) of **D-38**.

5.5   RELIEF OF OVER-CONSTRAINT

Steps (b) and (d) of **D-38** aim to relieve over-constraint inherent within a model graph by eliminating nodes in overlapping BNS. In their current form, the procedures for these steps state:

b) If overlapping BNS are found, eliminate nodes through negotiation so that all remaining BNS become non-overlapping.

d) If any two of these resultant constraint domains overlap, determine whether any knots are over-constrained. If so, by negotiation, remove sufficient nodes to relieve the over-constraint.

However, the removal of relations in their entirety from a model could be a bit offensive to the stakeholders who have contributed the relations to the model. This contribution represents their voice, their "*piece of the pie*", into the system design process. The exclusion of such input might not be practical or feasible in some instances, especially when this is the voice of the resource sponsor or policy maker who funds the project.

A less dramatic approach, and perhaps a more productive negotiation, to achieve buy-in from the affected stakeholder(s), would be to adjust some relations, rather than a total elimination of them. One possible technique would be to parameterize some constant coefficients to variables. Mathematically, the effect on p(**G**) should be identical, whether we remove a node from, or add a knot to a model graph. Such an adjustment technique, previously discussed in Section 1.3, could be employed as an alternative to steps (b) and/or (d) above.

Let's examine the example BPG of Figure 5-8, an inherently inconsistent model with two indirectly overlapping BNS. By adding a knot to the over-constrained node 5, the model now becomes inherently consistent. However, no computational request can be made on this extended model yet because it is perfectly constrained, consisting of two intrinsic 2×2 BNS and one resultant 1×1 BNS. In order to entertain any computational request, another knot needs be added to the BPG.

The above technique of extending adjustment to a BPG does not only allow more flexibility in system modeling and simulation for the model builder, but also facilitates thoughtful inquiry and cooperation among the systems engineering and program management team to expand the trade space in systems design.

5.6 EXPANSION OF RESULTANT CONSTRAINT DOMAINS

After all inherent over-constraints have been resolved by the model builder (with the aid of above method to detect directly overlapping BNS), those "variables" identified as parts of an intrinsic BNS have essentially become internal point-constraints per *T-11*. Their pre-determined values must now be propagated throughout the graph network to form *constraint domains*. This flow of constraint will fix other knots along the propagation path, and may even result in additional BNS. If two constraint domains share at least one common variable, they are said to be overlapping. In other words, the commonly-shared variable(s) are over-constrained. As illustrated in Figure 4-7, this phenomenon has been briefly discussed. As such, a constraint domain is now formally defined.

Definition 44: Let **G** be a model graph, and **H** be an intrinsic BNS within **G**, i.e. **H** is a subgraph of **G**. The ***constraint domain*** emanated from **H** is the maximum knot-set that includes K_H and all other knots in K_G which are also fixed as a result of constraint propagated from K_H throughout **G**.

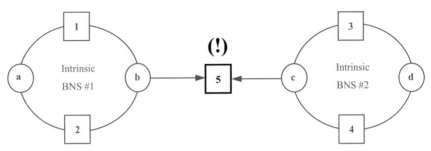

(a) Inherently inconsistent BPG with two intrinsic BNS over-constraining node 5.

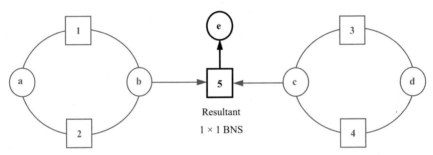

(b) Addition of knot e relieves over-constraint and renders BPG inherently consistent

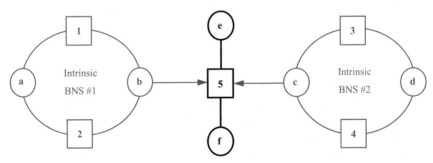

(c) Addition of knot f allows some computational requests made on BPG.

Figure 5-8. Adding knots to BPG relieves over-constraint and allows for computational requests

It must be emphasized that constraints may flow out of a BNS even further than the Duality Rules (*T-10*) imply. Even though the Duality Rules only work around tree-like structures, constraint can propagate through both tree-like and circuit-like structures. In tree-like structures, the Duality Rules can be applied very rapidly, but they will likely bog down in the vicinity of circuit structures where the BNS are hiding. It is quite possible that once a constraint domain is expanded as far as the Duality Rules allow, new resultant BNS may be formed. With all the knots reached by the Duality Rules now fixed (no longer "variables"), the remainder matrix can be used to search for more resultant BNS. Subsequently, the Duality Rules can be employed again to re-start constraint propagation from the newly discovered, and validated, resultant BNS. This expanded process, beyond the Duality Rules, has been termed the *Goldilocks Rule* to manage constraint flow through both tree-like and circuit-like structures. A constraint domain can certainly expand even beyond the circuit cluster in which the BNS reside, as far as its parent connected component will allow.

Figure 5-9 illustrates a model graph with several constraint domains. On the left-hand side, constraint flows from three intrinsic BNS into neighboring sub-graphs, which results in two additional BNS. The path of constraint propagation runs along the bold edges. On the right-hand side, each resultant constraint domain is shown to consist of its intrinsic BNS and the maximum sub-graph expanded by propagating constraint from the intrinsic BNS. Note that *a constraint domain may not only contain BNS, which are simple circuits or unions of adjacent circuits, but also tree structures.* This is the case with constraint domain #2 below. All variables within a constraint domain have been pre-determined (fixed) even before any computational request is made on the model.

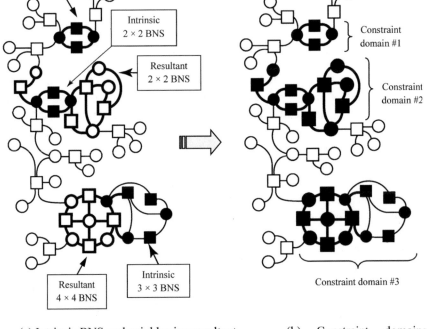

(a) Intrinsic BNS and neighboring resultant   (b)   Constraint   domains
    BNS within model graph.

Figure 5-9. Model graph with intrinsic BNS and resultant constraint
domains

Also note that any interface between a sub-graph which represents a
constraint domain and the remainder of a model graph is through the knot set
of the sub-graph. In other words, a constraint domain can be considered as a
maximum complete sub-model all of whose variables have been inherently
fixed. To expand the resultant constraint domain within a connected graph,
the following algorithm refines steps (c) – (f) of **D-38** with more details.

Definition 45: Procedure to expand the constraint domain $\mathbf{K_{cd}}$ from a
BNS within a <u>connected</u> graph **G**.

1) Per step (c) of **D-38**, propagate constraint from each and every knot of
   the source BNS by employing the Duality Rules of **T-10** to the farthest
   extent possible within **G**.

a. For each knot discovered along the constraint flow path, add it to $\mathbf{K}_{cd}$.

b. For each vertex along the path, including the source BNS, remove it from $\mathbf{G}$ via a vertex-deletion operation (Phan, pp. 101-102). This may fragment the remainder graph into several disjoint connected sub-graphs as evidenced from Figure 5-9. As a quick check for this fragmentation, or disconnectivity, of the remainder graph, Friedman [1967, p. 91] has proven that "*a model graph with V vertices and more than* $\left(V^{2}/4\right) - V + 2$ *edges is connected*".

c. If any over-constraint encountered during propagation, then $\mathbf{G}$ is inherently inconsistent. Employ technique(s) outlined in Section 5.5 to relieve local over-constraints.

d. Re-process $\mathbf{G}$ through steps (a) – (h) of $\mathbf{D\text{-}37}$ to identify a new set of intrinsic BNS since the model topology may be unpredictably altered as a result of step (c). Also repeat steps (1a) – (1c) after the new set of intrinsic BNS has been validated.

2) Per step (e) of $\mathbf{D\text{-}38}$, locate potential resultant BNS, and possibly continue expanding the constraint domain further, by re-processing the remainder subgraph(s) through steps (a) – (h) of $\mathbf{D\text{-}37}$.

a. To facilitate computation, a vertex-deletion subgraph can be represented by a *remainder matrix* (see $\mathbf{Definition\ 46}$ below).

b. Re-apply the algorithm outlined in $\mathbf{Definition\ 43}$ to detect overlapping among newly located BNS and to assist the model builder to resolve any inherent over-constraint among them.

c. Similarly to steps (1c) and (1d) above, if $\mathbf{G}$ is altered to relieve any local over-constraint among the newly located BNS, it must be re-processed through $\mathbf{D\text{-}37}$.

d. Resume propagation of constraint from resultant BNS by re-iterating steps (1) – (2c) until no more BNS can be located. As such, $\mathbf{K}_{cd}$ has reached its maximum and cannot expand any farther.

3) Repeat steps (1) – (2) for each and every intrinsic BNS within $\mathbf{G}$ to define its respective constraint domain.

4) Per step (d) of $\mathbf{D\text{-}38}$, re-apply the algorithm outlined in $\mathbf{Definition\ 43}$ to detect overlapping among the constraint domains defined in step (3).

a. If any overlapping detected, then $\mathbf{G}$ is inherently inconsistent. Employ technique(s) outlined in Section 5.5 to relieve over-constraints.

  b. If **G** is altered to relieve any overlapping among the constraint
      domains, it must be re-processed through steps (a) – (h) of **D-37**.
      Also re-iterate steps (1a) – (4a) above.

5) By step (f) of **D-38**, upon successful completion of step (4b) without any
   more overlapping among the BNS or among their resultant constraint
   domains, **G** can be concluded as an *inherently consistent* model graph.

Definition 46: The **remainder matrix** representing a vertex-deletion
subgraph **H**, denoted as X_H, of a model graph **G** is defined as the constraint
matrix [C_G] less the rows and columns corresponding to the nodes and
knots, respectively, which have been fixed within a constraint domain and
thus removed from **G**.

Figure 5-10 graphically demonstrates an application of the high-level
algorithm outlined in **Definition 45** above.
  • After the "kissing" circuits and circuit cluster of original model
    graph **G** have been separated at knots b and d by step (f) of **D-37**,
    intrinsic BNS_1 = { 1, a, 2, b } and BNS_2 = { 7, d, 8, g } can be
    identified by step (f).
  • Upon removal of BNS_1 and BNS_2 from **G**, the **D-37** process is re-
    applied to vertex-deletion subgraph **H**. Again, its "kissing" circuits
    are separated at knot f via step (f) of **D-37**. And this separation
    results in BNS_3 = { 3, c, 4, f } and BNS_4 = { 5, e, 6, f } by step (g) of
    D-37.
  • Note that BNS_3 and BNS_4 over-constrain knot f, which can be
    subsequently detected by applying the algorithm in **Definition 43**.

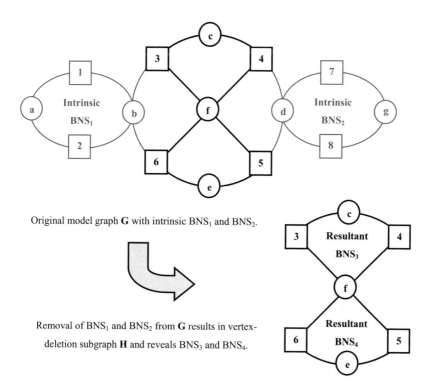

Original model graph **G** with intrinsic BNS₁ and BNS₂.

Removal of BNS₁ and BNS₂ from **G** results in vertex-
deletion subgraph **H** and reveals BNS₃ and BNS₄.

Figure 5-10. Removal of intrinsic BNS and repeating **D-37** locate resultant
BNS

Step by step, Figure 5-11 below demonstrates another application of the
algorithm in **Definition 45** by employing the example BPG of Figure 5-8:

• The flow of constraint is started from knot b of the intrinsic BNS, as the
source of internal constraint, into node n_5. The vertices of this source
BNS are removed from the model graph via a vertex-deletion operation
(Phan, pp. 101-102) to simplify the graph during propagation.

• By the **T-10** rule of $d(n) - 1$ in and 1 out for nodes, constraint continues
to flow from n_5 into knot c, after which n_5 is removed to further reduce
the graph.

• By the **T-10** rule of 1 in and $d(k) - 1$ out for knots, constraint can
simultaneously flow from c into n_3 and n_4. However, as a matter of
practical implementation, the branches of propagation needs be executed
one of a time, in a single thread of execution. Parallel threads of
execution for multiple branches may result in various issues with
concurrency, e.g. unpredictable racing condition. And it would be

difficult, if not impossible, to maintain (configuration control over) the state of the remainder matrix when it may be modified by several threads of execution. As such, for the purpose of demonstration herein, we will just trace through one branch of constraint flow from c to n_3.

- As knot c is removed to further simplify the graph, the vertex-deletion operation stipulates that all edges incident upon c must also be removed. And this includes the edge between c and n_4 (Phan, pp. 101-102).

- Again, by the **T-10** rule of $d(n) - 1$ in and 1 out for nodes, constraint continues to flow from n_3 into knot d, after which n_3 is removed to reduce the graph even further.

- Again, by the **T-10** rule of 1 in and $d(k) - 1$ out for knots, constraint can propagate from d to n_4.

- Without any outlet from n_4, the flow of constraint stalls here. Remember that the edge between n_4 and c has already been deleted above, and no longer exists in the remainder matrix. This violates the **T-10** rule of $d(n) - 1$ in and 1 out for nodes. Therefore, the original model can be concluded as inherently inconsistent, and no computational requests can be made on it.

The procedure outlined in **Definition 45** and demonstrated above should have better efficiency in terms of both run-time and space-bound than steps (c) – (f) of **D-38** in their current form. The removal of every vertex discovered along the constraint propagation path can repeatedly simplify the model graph, and thus reduce the size of the remainder matrix further and further. As such, the computational load for subsequent iterations will require less and less CPU time and memory resources.

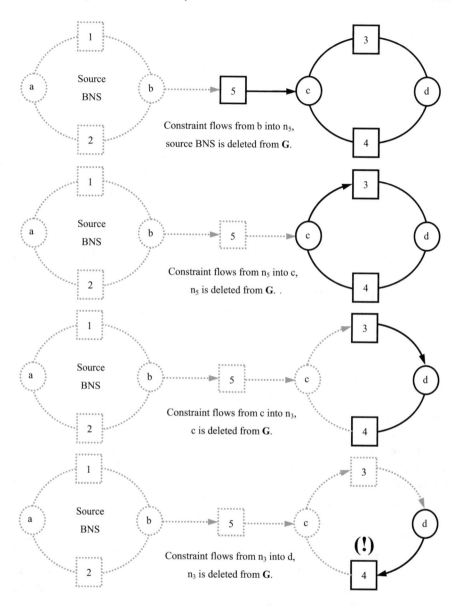

Figure 5-11. Violation of the d(n) – 1 in and 1 out rule for nodes exposes an inherently inconsistent model

5.7 PROCESSING OF COMPUTATIONAL REQUESTS

"As we observed from Chapter 1 and even more in Appendix A, even though the model may be consistent, the probability that any given computational request will be allowable is still very small," remarked the manager. "Can constraint theory assist the math model manager in steering him to those computational requests which *are* allowable?"

"Certainly," the analyst assured. "We can consider another 'brute force' approach here, where constraint theory can theoretically examine all the possible 2^K computational requests. But, as we have already seen, 2^N computations are really out of the question and 2^K would be roughly the same size."

"We can enormously reduce the number of computational requests to be analyzed by noting two things," the analyst continued. "First, the cognitive limitations of any human attempting to understand a result from a math model are limited to 'view spaces' whose dimensionality ranges from 2 to 5 at the most. So the 2^K computations are reduced to $\sim K^5$ – a tremendous reduction. Second, not all the K variables are equally interesting or important; for large models, probably only a few percent of the total number of variables would be considered as providing a 'system level' overview. We can take a hint as to which are the most interesting (or valuable) variables by just looking at the computational requests which were disallowed. Assume for example that a failed computational request involved the variables a,c,f,m. We could then ask D38 to examine all combinations of three of these and any other of the K – 4 variables, and two of these and any two other of the K – 4 variables, and finally any one of these and any three other of the K – 4 variables. This amounts to less than $4K^3$ examinations, which is over a million possible computational requests but can be computed by D38 in just a few minutes. This should provide the manager with many computational requests which are in the 'vicinity' of the one he wanted and couldn't have."

Given an inherently consistent model graph by steps (a) – (f) of *D-38*, we are now ready to examine in further details the general procedure for processing computational requests as outlined in steps (g) – (k).

5.7.1 INITIAL SIMPLIFICATION OF MODEL GRAPH

We can again leverage *T-11* to simplify a model graph before entertaining any computational request made upon it. Since all the knots belonging to a resultant constraint domain, previously identified by the method outlined in *Definition 45*, have been internally fixed in values, they can no longer be manipulated as part of any computational request. Therefore, the sub-graphs spanning the various constraint domains can be removed from the overall model, via a vertex-deletion operation (Phan, pp. 101-102) after step (f) and before step (g) of **D-38**.

This reduction can be safely implemented without losing any essential information in determining the allowability of any computational request, or the processing of an allowable request. Within the framework of such a reduced graph, one can avoid many repetitious and unnecessary calculations. And computational complexity can be exponentially lessened in terms of both run-time and memory-space requirements. This improvement in efficiency follows the same concept as presented in step (1b) of **Definition 45** where vertices discovered along a constraint flow path are removed during propagation of internal constraint.

Figure 5-12 illustrates above simplification technique. Upon the removal of the intrinsic BNS = { 2, 3, 4, c, d, s } which just happens to be the same as its resultant constraint domain in this case, model graph **G** is reduced to sub-graph **H**. All computational requests made on **G** can be evaluated with the remainder matrix associated with **H**. Any request involving c, d or s, as dependent or independent variables, can be readily concluded as unallowable since their corresponding knots $\notin \mathbf{K_H}$. This immediate determination will help avoid the unnecessary propagation of constraint as stipulated by steps (h) – (j) of **D-38**. Allowability of other requests, as well as the processing of all allowable requests, can also be computed more efficiently in **H** than in **G** by the fact that $|\mathbf{V_H}| \leq |\mathbf{V_G}|$ and $|\mathbf{E_H}| \leq |\mathbf{E_G}|$.

By **T-7**, the number of possible computational requests on a model with K variables is equal to 2^K. In the above simple example BPG, with $|\mathbf{K_G}| = 8$, there would have been $2^8 = 256$ computational requests that could be possibly made on **G**. With $|\mathbf{K_H}| = 5$, however, the number of possible computational requests on **H** is now only $2^5 = 32$, an order of magnitude less than that on **G**. For real-world models involving hundreds, if not thousands, of variables, the reduction in computational complexity would be even more significant.

Another advantage of the above simplification technique is that the removal of sub-graphs spanning various constraint domains may also break up a BPG into several disjoint connected components. In such cases, a computational request involving variables not all of which belongs to the same connected component can also be quickly determined as unallowable without having to propagate needlessly throughout the graph network. This is in accordance with *T-5*.

5.7.2   SIMPLIFYING MODEL GRAPH DURING CONSTRAINT PROPAGATION

To evaluate allowability of a computational request made on a model, steps (h) and (i) of *D-38* repeatedly apply the Duality Rules of *T-10* and the BNS search process of *D-37* to propagate constraint throughout its graph network. In their current form, the general procedures for these steps state:

h) Propagate computational paths from all constraint sources throughout the model employing the *T-10* rules. If any knot is over-constrained, the computational request is not allowable.

i) If the computational path does not reach the dependent variable then examine the residue for BNS, using *D-37*, and continue with *T-10*, followed by *D-37* as necessary until the computational path can go no further.

For computational requests with multiple independent variables as input, the processing should initiate the propagation of constraint externally applied to one independent variable at a time in a single thread of execution, and complete the *D-38* procedure as far as *T-10* and *D-37* will allow before starting propagation of constraint from another independent variable. To maintain flow control, independent variables should be processed sequentially one at a time. If several independent variables are processed in parallel, their computational paths may unknowingly collide with unpredictable results. Given that, as vertices are discovered one by one along the computational path originated from one independent variable, they can be dynamically removed from the BPG by a vertex-deletion operation (Phan, pp. 101-102). In several ways, this reduction can simplify computational complexity, and thus improve efficiency of the iterative process stipulated by steps (h) and (i) of *D-38*. The operational benefits include:

- With the ever-reduced remainder matrix, the *D-37* process to locate resultant BNS can only execute faster and faster, and require less and less computer memory resource with each subsequent iteration, than processing the entire model graph each and every time. The same can also be said of the *T-10* process to propagate constraint.

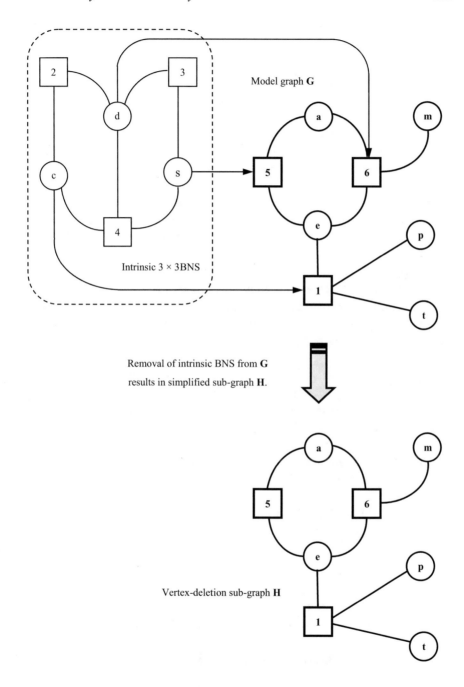

Figure 5-12. Removal of BNS simplifies a model graph before processing any computational request

- Deletion of vertices may break up the BPG into several disjoint connected components. Before initiating another path of propagation from the next independent variable, a quick and easy check can be performed to verify this independent variable as belonging to the same component as the dependent variable. If not, it can be readily concluded that the computational request is not allowable per *T-5*. This immediate determination will help avoid the repetitive and unnecessary propagation of constraint from this independent variable, as well as others subsequently.

- If the computational path initiated from one independent variable discovers the knot associated with another independent variable, then the request can be immediately rendered as unallowable.

Theorem 30: If the constraint flow path originated from an independent variable x_1 as input to a computational request discovers, by the Duality Rules of *T-10* for propagating constraint throughout a graph network, the knot associated with another distinct independent variable x_2 of the same request, then the request is <u>not</u> allowable.

Proof: Independent variable x_2, as input to the computational request just as x_1, will also be externally constrained by definition. If x_2 lies on the computational path originated from x_1 then x_2 is over-constrained once the flow of constraint from x_1 flows into it. This violates the propagation rule of "*1 in and d(k) – 1 out*" for knots under *T-10*. Over-constraint on x_2 causes the model to become inconsistent. And no computational request on an inconsistent model is allowable per *T-1*.

5.7.3   UNALLOWABLE COMPUTATIONAL REQUESTS

Steps (h) and (j) of *D-38* have pointed out two possible scenarios in which a computational request becomes unallowable due to over-constraint of a <u>knot</u> along the computational path. The following sub-sections will discuss other classes of unallowable computational requests.

5.7.3.1   OVER-CONSTRAINT OF NODES

For a given computational request, there may exist multiple possible computational paths, depending on the order (or sequence) of propagating externally-applied constraints. Per *T-10*, a node along a computational path will become over-constrained (or under-constrained) if the propagating rule of "*d(n)* – *1 in and 1 out*" for nodes is violated. Accordingly, such a computational request may also be rendered as unallowable.

Figure 5-1 graphically illustrates this situation. If the two constants in the bottom half of the graph are propagated first, then knots d and g will be discovered as over-constrained as stipulated by steps (h) and (j) of *D-38*. At this time, the procedure can be halted and the computational request declared unallowable. However, if the two independent variables in the top half of the graph are applied first as external constraints, then node a will be discovered as over-constrained. Accordingly, the same computational request can also be rendered unallowable and the procedure immediately halted.

5.7.3.2 RELEVANCY OF DEPENDENT VARIABLES

Step (k) of *D-38* states that: "If the path reaches the dependent variable by employing all independent variables, without either over- or under-constraint along the way, then the computational request is allowable. It is acceptable to have local under-constraint elsewhere in the model. If the dependent variable can be reached without having to propagate the constraint externally imposed on one or more independent variables, then the computation is unallowable".

Per the following concepts in Constraint Theory, another class of unallowable computational requests involves extra independent variables whose externally-applied constraints are not necessarily used to reach the desired dependent variable.

Definition 9: (p. 34): "y is a relevant variable with respect to relations ϕ in xyz space means that there exist lines in xyz space parallel to the y-axis that are neither entirely within nor entirely outside of the relation set. Thus, y has an effect on ϕ, or equivalently, the relation p constrains y".

Definition 14: (p. 40): "a computational request on a model is allowable means that the projection of A_Σ onto the view space of the computation contains at least one point and, in addition, each variable involved in the computation must be relevant to this projection in the sense of Definition 9".

It has been previously illuminated and asserted (pp. 40, 45) that: "If the projection has variables that are not relevant, these variables take on all their possible values, and are therefore under-constrained. Allowability requires that all the variables of the requested computation be relevant to the projection of the total relation onto the computational sub-space."

Let's consider the computational request e = f₁(m, t) made on sub-graph **H** of Figure 5-12. Figure 5-13 illustrates the propagation of constraints externally imposed on the independent variables m and t (highlighted with double squares). As knot m is initially fixed, its constraint flows into node 6, resulting in a 2 × 2 BNS. In turn, this resultant BNS fixes knots a and e, and the computational request is thus completed. It was not necessary, or useful, to propagate the external constraint imposed on t. Had the constraint on t been propagated first (before that on m), the computational path would never pass node 1 to reach e. Violating the "*d(n) – 1 in, 1 out*" rule for propagating constraint through nodes, node 1 would have been under-constrained. Therefore, this computational request is not allowable since t is irrelevant in determining e. In other words, e can be sufficiently determined without t.

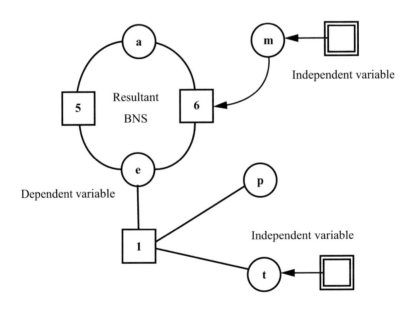

Figure 5-13. All independent variables must be relevant for a computational request to be allowable

Per **T-2**, over-constraint must not exist anywhere in the model, for a computational request to be allowable, since any sub-model inconsistency would "poison" the entire model. However, it is acceptable to have locally under-constrained vertices outside of the computational path. For example, in Figure 5-14, the computational request m = f₂(a) is allowable even though variables p and t are locally under-constrained elsewhere.

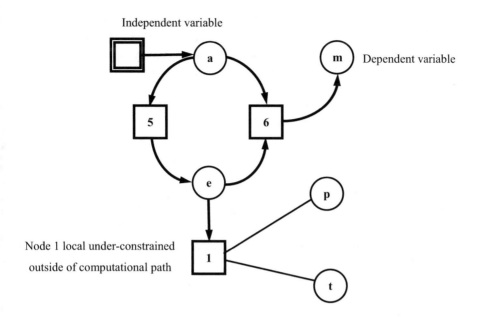

Independent variable

Dependent variable

Node 1 local under-constrained

outside of computational path

Figure 5-14. Locally under-constrained vertices outside
of the computational path are acceptable

5.8 SUMMARY OF CHAPTER AND CONSTRAINT THEORY TOOLKIT

The computational rules used in Chapter 1, "d(n) − 1 in and 1 out" for nodes and "1 in and d(k) − 1 out" for knots, as well as the "BNSs being the kernel of constraint" were formally sanctified from the viewpoint of the math model's being a set of relations within the multidimensional space defined by the model's variables. All these results were seen to be aspects of the Goldilocks rule which stated that computational allowability requires that the resultant constraint along the entire computational path from independent to dependent variables be exactly zero.

The exhaustive search for BNSs requires 2^N examinations of math model relation subsets; even for moderate model sizes of N = 100, thousands of universe lifetimes are required, even for nanosecond examinations.

Instead, *D-37* can locate the BNSs in only seconds instead of trillions of years. This is done by analyzing the topology of the BNSs which are imbedded within the topology of the bipartite graph meta-model, employing easily computed features such as connectedness, tree-ness, circuit rank and constraint potential. A key result is the proof that, if there is a BNS within a bipartite graph, it can only exist within a circuit cluster (cc) and if the constraint potential of that circuit cluster p(cc) is equal or greater than zero, then that circuit cluster contains p(cc) + 1 BNSs. See Figure 5-15.

Once the BNSs are located, the model consistency and computational allowability are easily determined, as is summarized in Tables 5-1 and 5-2, the Constraint Theory Toolkit.

Figure 5-16 demonstrates a typical scenario for the flow of constraint across a small bipartite graph model.

The many definitions and theorems of Chapters 4 and 5 may appear onerous but they are necessary to establish the rules of *D-37* and *D-38* as applicable to *any* mathematical model of *any* size. This is a demonstration of the power of generalizability of mathematics. We start in tiny domains of low dimension which we can comprehend, and then extend our understanding and tools to dimensions of any size. As large as the numbers 2^N and 2^K are, the number of possible topologies of a bipartite graph is unimaginably larger: 2^{NK}.

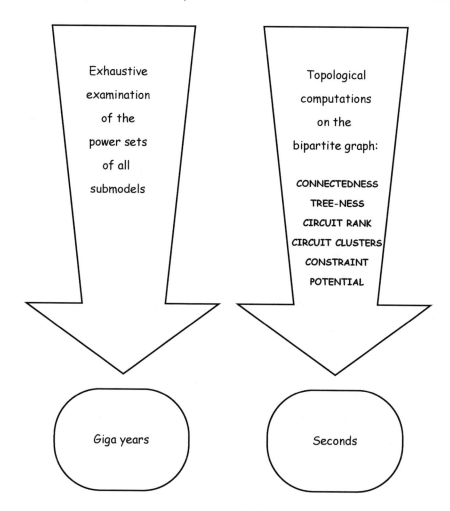

Figure 5-15. By employing easily computed topological properties of a model's bipartite graph, consistency and allowability checks can be reduced from universe lifetimes to seconds.

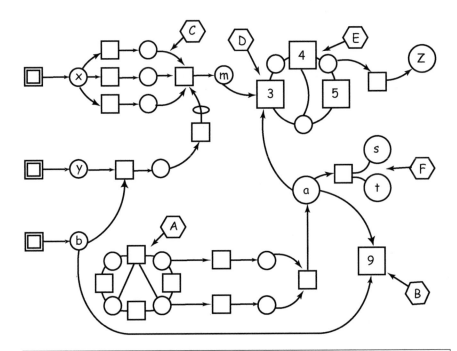

Notes:

A: This is a BNS, providing an intrinsic source of constraint to all its knots.

B: Constraint flows from the BNS to knot a, which coupled with the
       independent variable, b, serves to overconstrain node 9. This node
       must be removed in order to render the model consistent.

C: Although this is a circuit cluster, it has no BNS since p(CC)<0.

D: Using the rules for constraint flow in trees, the flow progresses from
       independent variables and the BNS to node 3 where it appears to halt
       because node 3 has four independent variables and only two are
       furnished as inputs.

E: However, the circuit cluster 345 becomes a resultant BNS since its
       constraint potential was increased from -2 to zero because of the two
       inputs from m and a. This permits the flow to continue to the dependent
       variable z, proving that the computational request, z=f(x,y,b) is allowable
       (providing node 9 is removed of course).

F: The constraint flow does not reach the knots s and t; they are
       underconstrained and would cause unallowable computations.

Figure 5-16. A walk through the trees and tangled clusters.

Table 5-1. Constraint Theory Toolkit (Part I).

To Determine Mathematical Model Consistency:

Organize the set of regular relations and their relevant variables into a bipartite graph (BPG) metamodel with two types of vertices:

- Nodes (N) which represent relations, and Knots (K) which represent variables
- Edges connect relations (nodes) to their relevant variables (knots); a BPG with undirected edges represent the math model; a BPG with directed edges represents the computation
- The degree of a vertex, d(v) is the number of edges which intersect that vertex.

Employ the companion Constraint Matrix (CM) to communicate with computer analysis.

The constraint propagation rules across a BPG are:

- for nodes: d(n)-1 edges flowing into the node, 1 edge flowing out.
- for knots: 1 edge flowing into the knot, d(k)-1 edges flowing out.

Propagate connectivity along BPG edges (or CM dots) to determine connected components.

- If K+N-E=1 in a connected component, it is tree-like; no intrinsic constraint in component.
- If not tree-like, circuit rank=E-K-N+1=number of independent circuits in component.
- If the constraint potential, p(SM), of a submodel = N-K>0, intrinsic constraint exists there.

A submodel wherein N=K is a nodal square (NS); a nodal square with no NS within it is a basic nodal square (BNS) which is the kernel of constraint in the math model.

- Overlapping BNSs (with p(BNSs)>0) indicate overconstraint causing INCONSISTENCY.
- Constraint propagates from the BNSs to resultant constraint domains via the above rules.
- If resultant constraint domains overlap, they will either be redundant or INCONSISTENT.
- Inconsistency must be negotiated by the human model builders to resolve overconstraint.

BNSs can only exist within circuit clusters (cc's).

A systematic search for BNSs involves separating the connected components, trimming external trees, eliminating internal trees, separating kissing cc's by removing separating vertices, and finally computing the constraint potential of the remaining cc's.

Table 5-2. Constraint Theory Toolkit (Part II)

To determine computational allowability:

If a mathematical model is inconsistent, no computations on it are allowable.

Add the sources of extrinsic constraint – independent variables and variables held constant – to the sources of intrinsic constraint and propagate constraint through the BPG using the propagation rules above. If overconstraint occurs at any vertex, the computation is NOT ALLOWABLE. If the constraint does not propagate to the dependent variable, using the above procedure, search for BNSs which can continue the propagation. If the constraint flow still does not reach the dependent variable, the computation is NOT ALLOWABLE due to underconstraint. If neither over- or underconstraint occurs, the computation is ALLOWABLE.

5.9   QUERIES FOR THE REGULAR STUDENT

1. For the mathematical model of Problem 4.1, which of these computational requests are allowable? If not, why not?
   $g = f(r, c)$, $g = f(s, t)$, $m = f(w, h, s, t)$, $h = f(r, w)$, $d = f(g, r)$

2. Is the mathematical model of Problem 4.2 consistent? If so, develop at least two computational requests on this model.

3. Is the mathematical model of Problem 4.3 consistent? Which of these computational requests are allowable? If not, why not?
   $n = f(r)$, $q = f(p)$, $m = f(n, q)$, $p = f(m, r)$, $m = f(p)$

4. Prove that the bipartite graph of a model with N relations and K variables has 2^{KN} different topologies.

5. Derive the number of possible computational requests can be made on a model with N relations and K variables if it were not for T6 and if the relations which formed the model were not counted as computational requests.

Chapter 6 DISCRETE AND INTERVAL RELATIONS

The diminished utility of metamodels

6.1 METAMODEL ISSUES AND PERSPECTIVES

"You recall that in Chapter Three, we defined three types of relations," said the analyst. "The most important of these types – from the standpoint of math modeling – is the regular relation and was treated in Chapter Four. In this chapter, we will look at the other types, called 'discrete' and 'interval'.

"This must have represented quite an intellectual leap, starting I presume from the findings of regular relations," the manager suggested.

"On the contrary," the analyst differed. "The original research in constraint theory actually started within the domain of discrete relations. It was stimulated by a paper on the solution of simultaneous equations in Boolean algebra, written by Antonin Svoboda [10], who was a member of Friedman's PhD committee. The use of the extension and projection operators in set theory was more easily visualized when only points – rather than multidimensional curves and surfaces – were involved. To generalize even further, we should remember that thinking about ordinals is more fundamental and closer to metamathematical language than thinking and operating on cardinals. Logic precedes analysis."

The perspectives and tools which we found to be useful for regular relations will now be examined for their utility regarding discrete and interval relations.

© Springer International Publishing AG 2017
G.J. Friedman, P. Phan, *Constraint Theory*, IFSR International Series on Systems Science and Engineering, DOI 10.1007/978-3-319-54792-3_6

6.2 THE GENERAL TAXONOMY AND PRIMARY PROPERTY OF DISCRETE RELATIONS

The discrete relation defined by D19 may appear in a variety of forms, as is displayed in Figure 6-1. Relation 1, a polynomial in a single variable is similar to a regular relation, yet it will point constrain its one-dimensional space at each of its roots. Relation 2 is an example of Diophantine equations which permit only integers as allowable solutions. Note that most of the equations in Chapter 4 – such as the definitions of constraint potential and circuit rank – are this type of equation. Relation 3 is described by a "truth table" which lists every point in its allowable space. Relation 4 is a logical or Boolean equation, wherein all the variables take on the values of *true* or *false,* or more compactly, 1 or 0, respectively. Finally, Relation 5 – represented by a matrix – can represent even more abstract mathematical forms such as bipartite graphs, via their companion matrices. In this sense, discrete relations can be considered as meta-metamodels for the graph theory described in Chapters 2, 3 and 4.

All these diverse characterizations of discrete relations can be covered by:

Theorem 31: Every discrete relation is an intrinsic source of point constraint with respect to each of its relevant variables.

Proof: By D19, in a discrete relation the intersection of any line with that relation is a point or set of points. By choosing that line to be, in turn, each axis of the relation's space, each of the variables is seen to be point constrained by that relation. *QED.*

Thus, the problem of finding sources of intrinsic point constraint and the search for BNSs which was given so much attention in Chapter 4 is almost trivial for discrete relations.

In pursuit of constraint theory's general goal of determining consistency and computational allowability, considerations of overconstraint and multidirectional constraint propagation, as well as computational flow rules will be treated in the following sections.

6.3 BOOLEAN RELATIONS

Boolean relations represent a significant portion of the class of discrete relations and can be described by logical equations, truth tables, Venn and Veitch diagrams, or by the complete mapping of allowable states in the hyperspace of relevant variables. Figure 6-2 presents a simple example of the logical equation: $A=B\overline{C}$. In words, this merely states that A is true (1) if B is true (1) and if C is not true (0). The truth table of Figure 6-2b provides the value of A for all possible combinations of B and C. The Veitch diagram

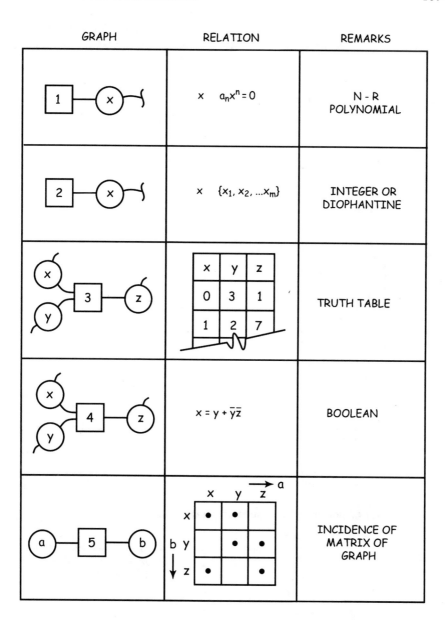

GRAPH	RELATION	REMARKS
1 — x	$x \quad a_n x^n = 0$	N - R POLYNOMIAL
2 — x	$x \quad \{x_1, x_2, ...x_m\}$	INTEGER OR DIOPHANTINE
x, y — 3 — z	x y z / 0 3 1 / 1 2 7	TRUTH TABLE
x, y — 4 — z	$x = y + \bar{y}\bar{z}$	BOOLEAN
a — 5 — b	incidence matrix x y z rows b x y z	INCIDENCE OF MATRIX OF GRAPH

Figure 6-1. Examples of Discrete Relations

– essentially a rectangular version of the Venn diagram – in Figure 6-2c also shows all possible values of the implied independent variables B and C and provides the value of A for each case. Sometimes, "forbidden domains" on the Veitch diagram are specified. These are combinations of the input variables which can never occur; when forbidden domains intersect domains defined by the logical equation, these forbidden domains take priority. This convention permits the simplification of logical equations which is important in the logical design of digital computers. In accordance with the hyperspatial viewpoints discussed in Chapter 2, when the total model is defined by a set of simultaneous Boolean equations, the *intersection* of the Boolean equations and the *union* of the forbidden areas are taken.

(a) Logical Equation : $A = B\bar{C}$

(b) Truth table

A	B	C
0	0	0
0	0	1
1	1	0
0	1	1

(c) Veitch Diagram :

A:

	C	$\bar{C}$
B	0	1
$\bar{B}$	0	0

(d) Set of points in hypersace of relevant variables :

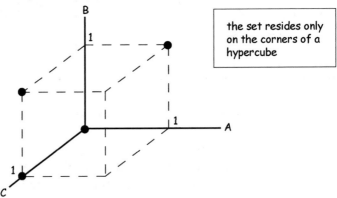

the set resides only on the corners of a hypercube

Figure 6-2. Boolean relations; and their representations.

Both the truth table and Veitch diagram representations suffer from implications of independent/variables and forbidden zone ambiguity.

Figure 6-2d displays the most complete representation of any discrete function: the identification of the set of points within the hyperspace of relevant variables. For any binary Boolean equation, this set can reside only at the corners of the hypercube formed by the points 0 and 1 along the axis of every relevant variable. That is, these are the values of the variables that are allowable by the defined relation. In the spirit of the four fold way presented in Chapter 2, call the set of points that satisfy the equation "A."

Let us now examine how the computational rule: (d-1)in/1 out that was developed for regular relations applies to Boolean relations.

Figure 6-3 displays all possible computational paths through the relation $A=B\overline{C}$. It shows that for B and C taken as input variables, every one of the four possible cases, the (d-1)in/1 out rule works. (This should not be surprising , since the format of the Boolean equation, truth table and Veitch diagram were predisposed towards this computational flow.) However, in the cases of A,B or A,C as input variables, it is seen that only half of the cases follow the (d-1)in/1 out rule, while the other half result in either multiple answers or represent forbidden inputs. Noting that multiple results also can occur in regular relations and can still be carried forward to other downstream computations, and that regular relations were defined to exclude forbidden inputs, it can be concluded that the (d-1)in/1 out rule *weakly applies* to discrete functions.

Figure 6-3 explores the more extreme case – from our regular relation perspective – of a single input variable. There exists one case out of six where the single variable A, set equal to 1, can uniquely determine the values of B and C. This is a case of 1 in/(d-1)out for nodes, which is surprising if we are used to thinking only of regular relations.

In summary, it can be seen that the computational flow across discrete relations depends far more on the precise nature of the relation than on metamodel rules such as (d-1)in/1 out.

Next, let us examine the applicability of constraint potential, (N-K), to determine the consistency of Boolean relations with K variables and N relations.

As noted above, a Boolean relation of K variables can exist only at the corners of a hypercube formed by the coordinates 0 and 1 along each of the K dimensions of the relation. Therefore, each of the N relations can be represented by a "binary number" containing 2^K states of A's for points contained by the relation and empty spaces which do not contain the relation.

Figure 6-4 provides the general format of a Boolean model. Note that, despite the fact that it is a rectangular array with the rows referring to variables and columns referring to relations, this is not a constraint matrix as defined in chapter 2. The total number of rows equals the number of possible states of any given Boolean function, and equals 2K, which is the power set of the K states. The total number of columns equals all possible Boolean functions which is the power set of the 2K possible states and equals $2^{(2^K)}$. In other words, the number of columns is the power set of the number of states which is the power set of the number of variables.

Checking computational flows

INPUT VARIABLES	OUTPUT VARIABLES	COMPUTATIONAL RESULT	
B, C = 0, 0			★
B, C = 0, 1	A	Unique result (from truth table)	★
B, C = 1, 0			★
B, C = 1, 1			★
A, B = 0, 0		C can be 0 or 1	
A, B = 0, 1	C	C is unique	★
A, B = 1, 0		input is forbidden	
A, B = 1, 1		C is unique	★
A, C = 0, 0		B is unique	★
A, C = 0, 1	B	B can be 0 or 1	
A, C = 1, 0		B is unique	★
A, C = 1, 1		input is forbidden	
A = 1	B, C	B, C are unique	★★
A = 0		B, C are ambiguous	
B = 1	A, C	A, C are ambiguous	
B = 0		A, C are ambiguous	
C = 1	A, B	A, B are ambiguous	
C = 0		A, B are ambiguous	

> ★   (d-1) in / 1 out rule works 8/12 times;
> ★★ 1 in / (d-1) out "rule"!

Figure 6-3. Computational Flows of A=B̄C

We now pose the question: If all the N columns refer to non-redundant Boolean equations, what is the largest N which will not cause the total model to become inconsistent? The answer to this question results in:

Theorem 32: For a 2-valued Boolean model consisting of K variables and N independent (non-redundant) relations, the largest N which will not overconstrain the model is: $N_{max} = 2^{2^{K}-1}$.

Proof: Recall that, from Chapter 2, the model will be inconsistent if the total allowability set is the null set. Also recall that the intersection of a null set with any other set results in a null set. By examining Figure 5-4 (b), we note that exactly half of the columns can be paired with the other half such that their intersections are null sets. (Specifically, look at columns 1 and 16, 2 and 15, …etc.) Also note that no other columns have null set intersections. Therefore the maximum number of relations which will not result in a null set is exactly half the total number of columns. In short, the maximum N which assures consistency is half the number of columns, or: $N_{max} = (\frac{1}{2})($

$2^{2^{K}}) = (2^{-1})(2^{2^{K}}) = 2^{2^{K}-1}$. *QED.*

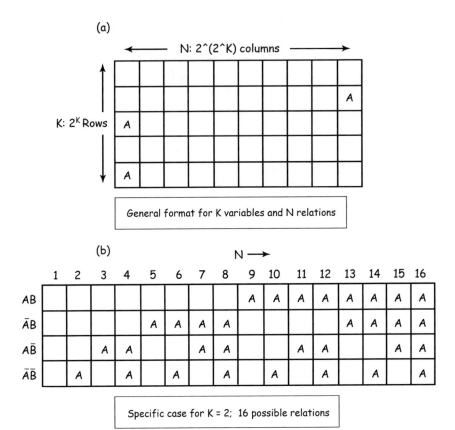

Figure 6-4. Proof of T32

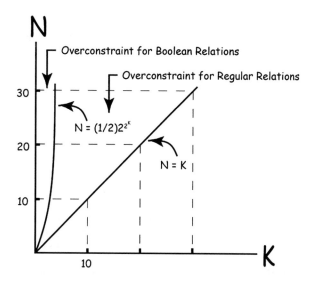

Figure 6-5. The Constraint Potential for Boolean relations is weaker than for regular relations.

Thus, the constraint potential (N-K) for regular relations can be replaced by $(N - 2^{2^{K}-1})$ for Boolean relations, and as is shown in Figure 6-5, it is a weaker indication of the model's inconsistency. Once again, we see that the specific properties of the relations must be examined and the metamodels, so useful in Chapters 4 and 5, are less useful for discrete relations.

Additional properties and methods of discrete relations are therefore discussed in the following.

6.4 TOPOLOGICAL IMPLICATIONS

These two types of discrete relations will have importance from a constraint standpoint:

Definition 47: A *full discrete relation* is an explicit discrete equation in which each point in the space of independent variables provides at least one value of the dependent variable.

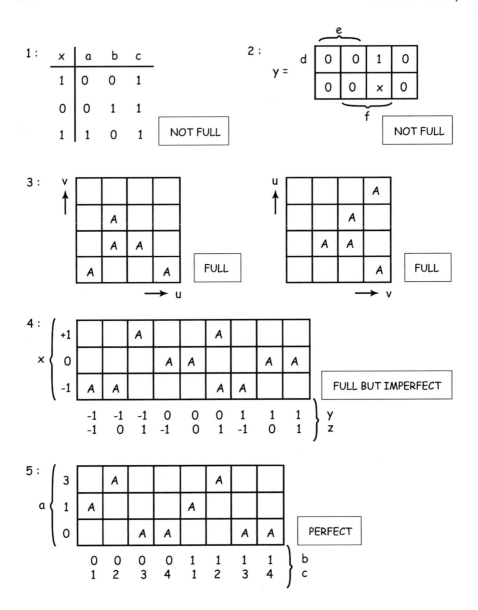

Figure 6-6. Examples of Full and Perfect Expressions.

Definition 48: A *perfect* discrete relation is a full discrete relation in which each point in the space of independent variables determines a unique value of the dependent variable.

Examples of these definitions are shown in Figure 6-6. Relation 1 is not full since only three of the possible eight points in the *abc*-space provide values for *x*. Relation 2 is not full because the point $\overline{def}$ is specified as being forbidden and thus has no output to *y*. The map, 3, relating to the 4-state variables u and v, is full if v is the dependent variable but is not full if u is made the dependent variable. Thus, "fullness" and "perfection" are properties of the explicit expression of a relation rather than of the relation itself. Relation 4 is full with respect to x but is not perfect because the "input" point $(y=0, z=1)$ yields two values for x: $=+1$ and -1. Finally, Relation 5 is perfect with respect to the ternary variable a because every point in bc-space determines a unique value for a.

Theorem 33: Every Boolean relation which is perfect with respect to at least one of its variables, is a universal relation.

Proof: Express the relation in its perfect, explicit form. By D48, every point in the product space of the input variables produces an output; thus the relation is universal. *QED*

Theorem 34: Every tree subgraph of Boolean relations each of which is perfect with respect to at least one variable, is consistent.

Proof: By T33, the tree has only universal relations, and by T9, this tree of universal relations is consistent. *QED*

Examples of the application of these Theorems are shown in Figure 6-7. In Figure 6-7a, the mere specification that Relations 1, 2 and 3 are Boolean, and that they are written in explicit form without specific forbidden domains allow us to conclude that they are perfect with respect – at least – to the dependent variables shown. By T34, then, the tree in Figure 6-7a is consistent. Note that the dependent variables of Relations 1, 2 and 3 do not indicate an obvious direction of computation or constraint flow.

Figure 6-7b presents a more general case of a discrete tree, mixing the binary variable y with the ternary variables v, w, x and z, and the quaternary variable r. Inspection of the truth tables for Relations 4, 5 and 6 discloses that each relation is universal. Thus, by T9, this discrete tree is consistent.

That these rules cannot be generally extended to submodels whose graphs are circuits is demonstrated in Figure 6-7c. Despite the fact that both of the Relations 7 and 8 are perfect with respect to all their variables, and both are universal, neither T9 nor T34 are applicable since they are connected in a circuit cluster. In fact, a plot on a klm-space Boolean map will show that the intersection of their relation sets is the null set for the given Relations 7 and 8, rendering the model inconsistent.

6.5 ALLOWABILITY OF DISCRETE COMPUTATIONS

A general approach to the determination of the allowability of any computational request made on discrete relations is described below.

The basic concept employed is a modification of the principles outlined by A. Svoboda [15] in solving simultaneous systems of Boolean equations.

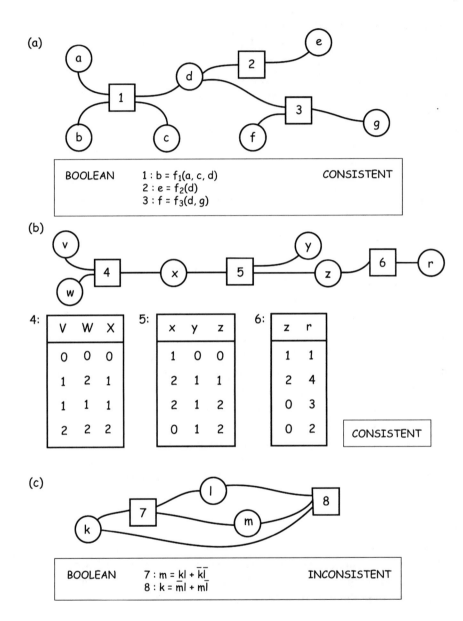

Figure 6-7. Trees of Discrete Relations.

Let us begin by applying a viewpoint established in Chapter 2. Refer to the top of Figure 6-8, where the discrete Relation 1 is shown in perfect form with respect to the 4-state variable z. Given this representation of the Relation 1, let us request the computation: $x=f(y,z)$, where x is binary and y

is ternary. By Figure 2-14, we must first take the *xyz*-space "view" of A_1 by taking the projection: $Pr_{xyz}A_1$. This is accomplished in Figure 6-8 by replotting the 12 points of A_1 from the *z* by *wxy* map to the *w* by *xyz* map, suppressing the value of variable *w*, and replotting in an *x* by *yz* map. Note that $Pr_{xyz}A_1$ contains only 11 points due to the fact that two points in A_1 project onto a single point, $(x,y,z)=(1,1,1)$ in the *xyz* view. The final step is to transform the *x* by *yz* map of $Pr_{xyz}A_1$ into a "function map" wherein the outputs for x are plotted as a function of the 12 possible points in the *yz* input space. As can be seen in the lower right hand corner of Figure 6-8, only 5 input points in *yz*-space yield unique results, while 3 input points yield multiple outputs and the remaining 4 yield no outputs at all; they are "forbidden."

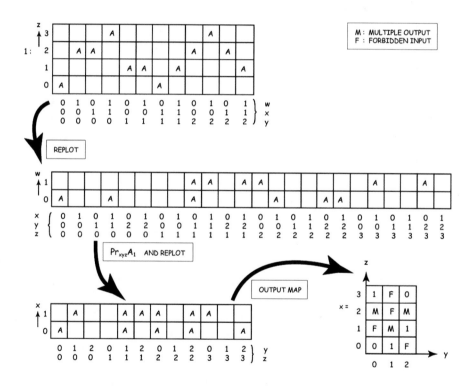

Figure 6-8. Determination of X=f(y,z) by projection and replot.

6.6 INEQUALITY RELATIONS

As defined by D19, an interval relation is any relation such that its intersection with a line is an interval containing an infinity of points.

For all practical cases, the interval relation may be represented by a system of inequality equations, or more generally by a combination of inequality and "equals or greater (less) than" equations.

Definition 49: *Inequality relations* are defined by "<" and ">". The *boundary* set of equations are formed by replacing each < and > with equal signs.

Figure 6-9 provides examples of D49.

a > b + d - e

p < m + q²

x + y > 3z

a) pure inequalities

a ≥ b + d - e

p ≤ m + q²

x + y ≥ 3z

b) equals or greater (less) than equations

a = b + d - e

p = m + q²

x + y = 3z

c) boundary set at equations for a) and b)

Figure 6-9. Comparisons between pure inequality equations, equals or greater (less) than relations, and boundary sets.

It was previously noted that, for regular relations, the BNS was required to provide intrinsic point constraint, and for discrete relations, intrinsic point constraint was always provided. We are now prepared to summarize the point constraint situation for all classes of relations:

Theorem 35: In discrete relations, intrinsic point constraint always occurs; in regular relations, intrinsic point constraint occurs in the presence of a BNS; in inequality relations, intrinsic point constraint never occurs.

Proof: The first portion is true by T31; the second is true by T11; the last is true because inequality relations do not include their boundary sets and thus all intersections either include an infinite number of points or the null set. *QED*

It is noteworthy that in a broad range of applications of math models, it is the boundary set which is of the greatest interest. An example of this is the general optimization techniques of linear programming where the optimization of a linear criterion occurs only at the intersection of the boundary sets of the constraining functions. Thus for these important applications, the inequality sets collapse into sets of regular relations.

The next consideration is to examine the utility of the constraint potential in determining consistency. As can be seen in Figure 6-10, the number of inequality constraints can increase indefinitely without driving the overall model into inconsistency. This observation permits us to state another generalization across all relation classes:

> ***Theorem 36:*** For model consistency, in discrete relations, $N_{max} = 2^{2^K-1}$; in regular relations, $N_{max} = K$; in inequality relations, there is no N_{max}.
> ***Proof:*** By T32, by T26, and by inspection of Figure 6-10. *QED*

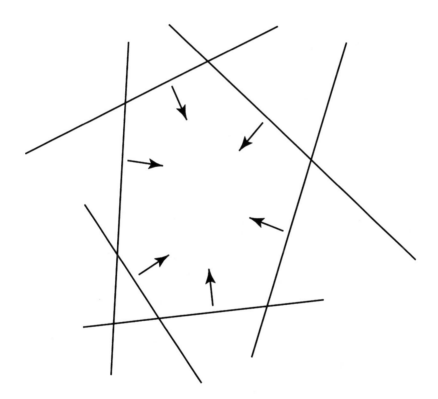

Figure 6-10. The number of inequality constraints can increase to infinity without driving the model's relation set to null (inconsistency).

Next, examining the propagation of constraint or computation through the metamodel, we invoke the transitivity rule of inequalities to prove:

> ***Theorem 37:*** If $a>b$, and $b>c$, then $a>c$.

As a side observation, if *a, b* and *c* are preferences, then the intransitive preference ordering of *a>b>c>a* is considered *irrational* by decision theorists.

Let us consider the implications of T37 on the allowability subset A in K-space. In the upper part of Figure 6-11, the "A in K-space" interpretation of the Theorem is presented. For ease of visualization, only the unit cube in the upper octant is shown. The two intersecting planes are the boundary sets for Relations 1 and 2 and the resultant allowability set $A_1 \cap A_2$ is the tetrahedron with the corners *o,x,y* and *z*.

The projection of $A_1 \cap A_2$ on the *ac* plane is seen to cover half the plane and corresponds to the interval relation *a>c*, substantiating T37. Thus we see that *a* and *c* are mutually constraining: constraint on *a* will propagate to *c* and vice versa.

On the other hand, consider the lower part of Figure 6-11, where the inequality *b>c* has been replaced by *b<c* rendering T37 inapplicable.

6.7   SUMMARY

• Although the general principles of constraint theory are still applicable to discrete and interval relations, the metamodel approach is less powerful than for regular relations. Therefore the metamodel viewpoints must be augmented by direct examination and computation of the full model itself.

• Certain generalizations can be made across all relation classes:

Regarding:	Discrete Relations	Regular Relations	Interval Relations
Intrinsic Point Constraint	At every node	At a BNS only	Never
Maximum N without overconstraint	$N_{max} = 2^{2^K - 1}$,	$N_{max} = K$	$N_{max} = $ infinite

The new allowability space becomes the pyramid with corners at *o,x,v,w,*and *z*. The projection of this pyramid onto the *ac* plane covers the entire plane and the variables *a* and *c* now do not exert constraint on one another, in the view defined in Chapter 2.

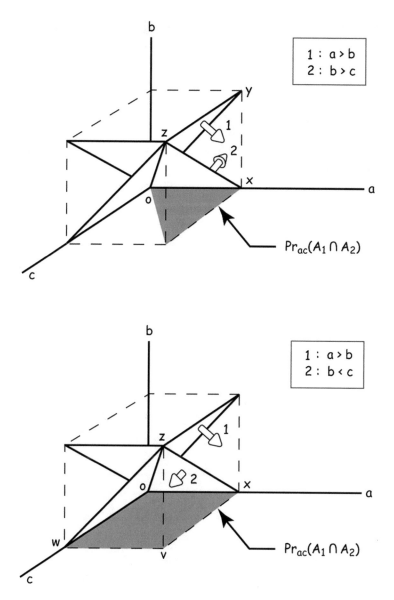

Figure 6-11. Transitive and intransitive inequality relations.

6.8   PROBLEMS FOR THE DISCRETE STUDENT

6.1 Given: a model of random Boolean relations with K variables and N relations. Derive: Probability of inconsistency as a function of K,N.

6.2 Construct a model comprised of three ternary discrete relations and determine under what circumstances the constraint flow follows the nodal rule: (d-1)in/1 out.

6.3 Construct two Boolean functions and find a third function which satisfy both.

6.4 Derive the general n-ary version of Theorem 32

Chapter 7 THE LOGICAL STRUCTURE OF CONSTRAINT THEORY

A Compact Summary

7.1   OVERVIEW

This chapter provides a summary of all the postulates, definitions and theorems presented in Chapters two through Six. Finally, two graph structures are presented, showing the interrelationships between the elements of constraint theory's logical structure.

7.2   POSTULATES AND PHILOSOPHICAL ASSUMPTIONS

Mathematical model building is a rapidly expanding activity. Since computers do not suffer the dimensionality limitations of the human mind, modeling is the best hope for systems engineers to manage the complexity of modern and future systems. All model builders and users share (or should share) these philosophical views and postulates:

Mathematics has been able to capture an incredible portion of natural and man-made phenomena with amazing depth and accuracy. Just as valuable is the ability of mathematics to logically integrate many diverse views of the world, generalizing new findings and helping to manage the vast dimensionality of complex systems.

© Springer International Publishing AG 2017
G.J. Friedman, P. Phan, *Constraint Theory*, IFSR International Series on Systems Science and Engineering, DOI 10.1007/978-3-319-54792-3_7

However, mathematics usually presents a simplified picture of the real world and the variables in its models may not be defined coherently across all members of a diverse team. These key issues are not within the domain of constraint theory and it is presumed that they have been attended to prior to the methods presented in this book.

Even when the issues of model accuracy and definitions have been handled perfectly, there still exist crucial issues of well-posedness. These are in the domain of constraint theory and the structure presented here addresses them. Specifically, are models consistent and are the computational requests made of them allowable?

Table 7-1 presents the postulates which have been directly involved in the development of constraint theory.

Table 7-1. Postulates

1. Model Builders inherently wish the relations in their model to be locally universal. (That is, if any relation applies a constraint to a variable, that variable will be able to propagate that constraint to an adjacent relation without causing inconsistency.)

2a. The laws of physics and other sciences, if they are fully understood, are fundamentally consistent.

2b. The intent of mathematical model builders is to represent phenomena accurately; thus if overconstraint – or inconsistency – occurs, it is unintentional.

7.3   DEFINITIONS

The forty definitions employed in Chapter Two through Six are listed in Table 7-2.

Most of these definitions either contribute to other, more complex definitions, or to theorems. In seven cases – marked by an asterisk – the definitions are a result of a set of theorems and represent a computational procedure.

7.4   THEOREMS

The thirty-three theorems derived in Chapters two through five are listed in Table 7-3.

The proofs of all these theorems – except T12, "Hall's Theorem" – are provided as they are first introduced in the chapters. The proof of Hall's theorem can be found in the reference.

Table 7-2. Definitions

D1: Set, Subset	D26: Adjacency
D2: Variable	* D27: Connectedness algr'm
D3: Model Hyperspace	* D28: Sep. Vertex Algorithm
D4: Product Set	* D29: Tree Algorithm
D5: Relation	* D30: Twig pruning algr'm
D6: Constraint	D31: Circuit Rank
D7: Union, Intersection	D32: Simple Circuit
D8: Projection, Extension	D33: Circuit Vector
D9: Relevance	* D34: Circuit Rank Algr'm
D10: Bipartite Graph	D35: Graph Taxonom
D11: Model Graph	D36: Independent BNS
D12: Constraint Matrix	* D37: BNS Location
D13: Consistency	* D38: General Procedure
D14: Allowability	D39: Overlap BNS
D15: Connected Component	D40: CharVec of BNS
D16: Tree Structure	D41: BNS Matrix
D17: Circuit Cluster	D42: Overlapping Factor
D18: Universal Relation	D43: Algorithm to detect
D19: Relation Classes	Overlapping BNS
D20: Locally Universal	D44: Constraint Domain
D21: Regular Relations	D45: Expanding Constraint
D22: Constraint Potential	Domain
D23: Degree of a Vertex	D46: Remainder Matrix
D24: Over Constraint	D47: Full Discrete Rel'n
Under Constraint	D48: Perfect Discrete
D25: Nodal Square	Relation
Basic Nodal Square	D49: Inequality Relations

7.5 GRAPHS OF THE LOGICAL STRUCTURE OF CONSTRAINT THEORY

The graphical portrayal of the relationships between definitions and theorems is presented in Figure 7-1. Although this graph has two disjoint sets of vertices, it is not a bipartite graph because in several cases, there are edges connecting definitions to other definitions, and edges connecting theorems to other theorems.

Another graphical portrayal is presented in Figure 7-2. This graph is a generalization of Figure 7-1 and suppresses many of the details but emphasizes the major logical thrusts of the theory. The more important definitions – such as the BNS – and the more important theorems – such as the BNS location theorem – are highlighted.

7.6   COMPLETENESS

The author makes no claim that these summaries and Constraint Theory are complete.

Indeed, it his hope that this work will stimulate further research and study into the increasingly important objective of managing complexity.

Table 7-3. Theorems

T1:   If a model is inconsistent, no computational requests on it are allowable

T2:   If any submodel is inconsistent, the total model is inconsistent

T3:   If two relations have no variables in common, they are consistent

T4:   If two disconnected components are internally consistent, they are consistent with each other

T5:   No computations across disconnected components are allowable

T6:   Allowability of a computational request is independent of the dependent variable

T7:   There exist 2^K possible computational request of a K-variable model

T8:   There exist 2^N possible submodels of an N-relation model

T9:   Any set of universal relations whose graph is a tree is consistent

T10:   Computational rules for models with a tree structure:
       nodes: d(n)-1 inputs --> 1 output
       knots: 1 input --> d(k)-1 output

T11:   Every BNS exerts point constraint on each of its variables

T12:   Hall's theorem

T13:   A subgraph is a tree if and only if V-E=1

T14:   The number of independent circuits in a graph = circuit rank of graph

T15:   The graph taxonomy of D35 is mutually exclusive and exhaustive

T16:   If SUM P_i > -n, then, at least one of the P_i > -1

T17:   If p(G)=N-K>0, G contains at least one BNS

T18:   No BNS can have a subgraph with a constraint potential>0

T19:   Every BNS must be connected

T20:   No BNS can be a tree

T21:   No BNS can have a tree which is not part of a circuit

(continued)

Table 7.3 Continued – Theorems

T22: No BNS can lie across circuit clusters containing a separating vertex

T23: No BNS can lie across a tree-like network of circuit clusters linked by a tree

T24: No BNS can lie across a tree-like network of circuit cluster linked to other circuit clusters by trees

T25: Every BNS is the union of adjacent circuits within a circuit cluster

T26: If a circuit cluster (cc) has p(cc) > 0, then it has at least p(cc)+1 BNSs

T27: The maximum number of BNSs in a $cc = 2^{c(cc)}$

T28: BNS_i and BNS_j are overlapping if $\omega_{i,j} \geq 1$

T29: Solution time to detect overlapping BNS and identify over-constrained variables using a matrix Ω_G with R rows and K columns $\sim K \cdot R^2$

T30: If the constraint flow path originated from an independent variable x_1 as input to a computational request discovers, by the Duality Rules of T-10 for propagating constraint throughout a graph network, the knot associated with another distinct independent variable x_2 of the same request, then the request is not allowable

T31: Every discrete relation is an intrinsic source of point constraint

T32: For a consistent model of Boolean relations, $N_{max} = 2^{2^K - 1}$

T33: Every Boolean relation which is perfect with respect to at least one of its variables is a universal relation

T34: Every tree subgraph of perfect Boolean relations is consistent

T35: In discrete relations, intrinsic constraint *always* occurs
In regular relations, intrinsic constraint *sometimes* occurs (in a BNS)
In interval relations, intrinsic constraint *never* occurs

T36: For consistency in discrete relations, $N_{max} = 2^{2^K - 1}$
For consistency in regular relations, $N_{max} = K$
For consistency in interval relations, $N_{max} = $ infinity

T37: If $a > b$, $b > c$ then $a > c$

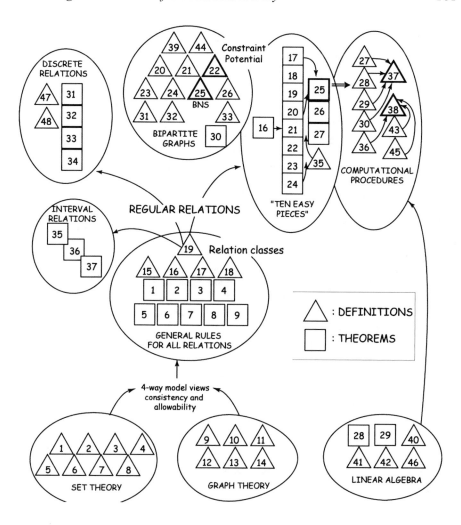

Figure 7-1. Graph of Constraint Theory Definitions and Theorems

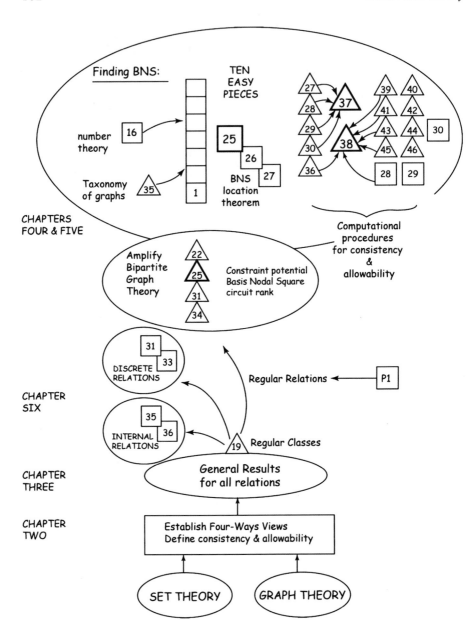

Figure 7-2. Logical Thrusts of Constraint Theory

Chapter 8 **EXAMPLES OF CONSTRAINT THEORY APPLIED TO REAL-WORLD PROBLEMS**

8.1   APOLOGIES NOT REQUIRED

The examples provided in this Chapter are drawn from applications in the aerospace industry that the author has experienced. Rather than apologize for this apparently undemocratic representation, it should be noted that it was the aerospace industry which gave the first major fertile ground for the discipline of systems engineering to manage the ever increasing complexity of integrating new technologies, equipment and missions.

The first example applies constraint theory to the operations analysis of new systems; it is an elaboration of the example presented in Chapter 1. The second example deals with the kinematics of free-fall weapons, which, despite the advent of "smart bombs" still has utility in many operational scenarios. The third example is a dynamic analysis of the control of the trajectory of an asteroid employing mass drivers.

8.2   COST AS AN INDEPENDENT VARIABLE (CAIV)

For several decades, the development process of new complex systems by the United States Department of Defense (DOD), has permitted the dominance of technical performance over cost and schedule. As a consequence, the vast majority of programs suffered major cost overruns and

© Springer International Publishing AG 2017
G.J. Friedman, P. Phan, *Constraint Theory*, IFSR International Series on Systems Science and Engineering, DOI 10.1007/978-3-319-54792-3_8

schedule slips. Essentially, performance was the primary driving requirement – or "independent variable." In order to achieve a greater measure of cost control and containment, the DOD initiated a management thrust called, "Affordability, or Cost as an independent variable (CAIV)". Figure 8-1 provides a briefing chart used by the US Air Force Research Laboratories (AFRL) that is intended to represent this new emphasis.

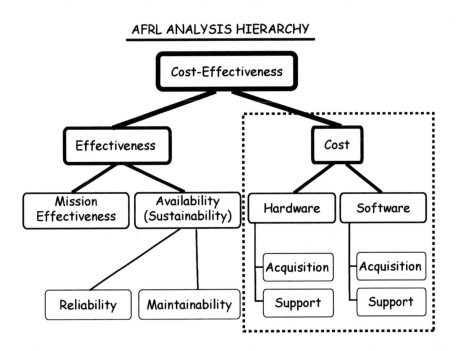

AFRL ANALYSIS HIERARCHY

**Affordability, or Cost As An Independent Variable (CAIV)
Is An Essential Component in Technological Trade Studies**

Figure 8-1. US Air Force Research Laboratories Briefing chart on Cost as an Independent Variable.

At this broad level of detail, the math model employed is similar to the example of Chapter 1. Mathematically, this type of problem is ideally suited to the viewpoint and tools of Constraint Theory. Simply stated, a model is built, and rather than follow the original computational flow from performance specifications to cost, the flow is reversed; the independent and dependent variables are switched. The new problem can be stated as:

"Given budgetary constraints on newly developed systems, what is the optimum system design which provides the most acceptable performance without exceeding these budgetary constraints?"

Let us flesh out in more detail the AFRL chart to attain a model which can perform the desired analysis. Table 8-1 presents a list of variables; note that K=14 across three levels of detail. Table 8-2 lists the relations, with N=16. Noting that N>K, we should be immediately concerned with model consistency.

Table 8-1. Variables for Model

DEFINITIONS	
CE	Cost Effectiveness
P_K	Probability of Target Destruction
A_V	Availability to Perform Mission
C	Total Program Cost
P_S	Weapon Survivability
CEP	Circular Error Probable
W/H	Warhead weight
$MTBF$	Mean Time Between Failures
$MTTR$	Mean Time to Repair
C_D	Cost of Development
C_P	Cost of Production
C_O	Cost of Operation
Wt	Total Weight
σ	Radar Cross Section

The top-level bipartite graph shown in Figure 8-2 is completely tree-like and has K=14 (representing all the variables) and N=5. Clearly, there is no concern about overconstraint or consistency at this level.

Table 8-2. Relations for Model

A "COHERENT" CRITERION MODEL			
1.	CE	=	$f_1 (P_K, A_V, P_S, C)$
2.	P_K	=	$f_2 (CEP, W|H)$
3.	A_V	=	$f_3 (MTBF, MTTR)$
4.	C	=	$f_4 (C_D, C_P, C_O)$
5.	P_S	=	$f_5 (Wt, \sigma)$
6.	MTBF	=	$f_6 (C_D)$
7.	σ	=	$f_7 (C_D)$
8.	Wt	=	$f_8 (W|H)$
9.	CEP	=	$f_9 (C_D)$
10.	C_P	=	$f_{10} (Wt)$
11.	C_O	=	$f_{11} (MTTR)$
12.	MTBF	$\geqslant$	K_1
13.	Wt	$\leqslant$	K_2
14.	σ	$\leqslant$	K_3
15.	C_P	$\leqslant$	K_4
16.	CEP	$\leqslant$	K_5

However, the 9 variables at the bottom of the graph are all inter-related, as is shown in Figure 8-3 where 6 additional nodes are added to indicate these relations. For example, node 7 describes how the cost of development, C_D, influences the radar signature, and node 9 describes how C_D influences the accuracy (CEP) of weapon delivery. Although we now have accumulated several circuits, the total constraint potential is still less than zero, and there are no local domains where the constraint potential exceeds zero – thus, there are no BNSs. This can be verified by examining Figure 8-5 which displays the constraint matrices for the various levels of detail in the model.

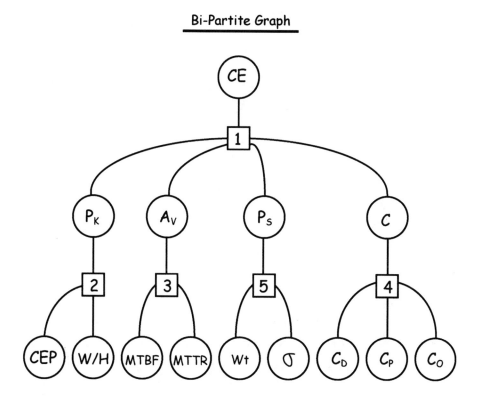

Bi-Partite Graph

Figure 8-2. Top-Level Bipartite Graph.

The constraint situation gets far more serious when the final 5 nodes at the bottom of Table 8-2 are included in the model. Note that each of these last five nodes are "policy demands" specifying additional requirements on reliability, weight, radar signature, etc., without directly relating them to a higher level system criterion. The addition of these is devastating to the consistency of the model. As is shown in Figure 8-4, the application of nodes

12, 14 and 16 drives C_D into inconsistency. Similarly, the application of nodes 10 and 15 drives C_P into inconsistency.

Thus, the bipartite graph and constraint matrix provide the managers and analysts visibility regarding the model's consistency, paving the way towards an eventual "cost as an independent variable" analysis. When overconstraint does occur, it can be pinpointed and – just as was done in Chapter 1 – allowing policy nodes such as 12-16 of Table 8-2 to be reconsidered for model inclusion. In short, rather than specifying a mandatory level of reliability, signature, etc., let these variables "run free" in the network of all other relations and variables. In this fashion, optimum values can be computed rather than dictated.

Bi-Partite Graph

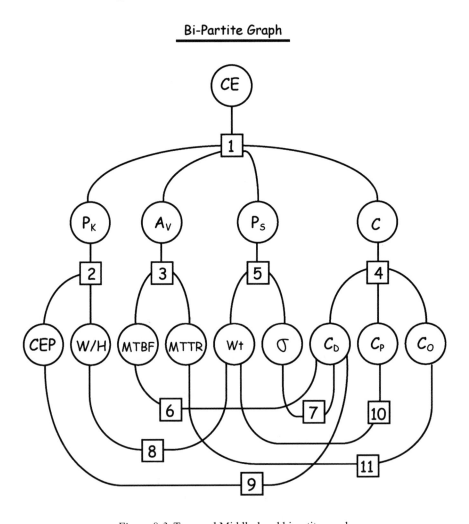

Figure 8-3. Top- and Middle-level bipartite graph.

Bi-Partite Graph

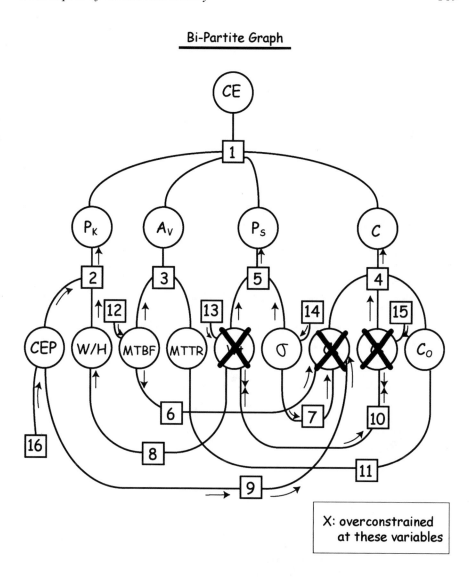

Figure 8-4. Full Bipartite Graph.

Constraint Matrix

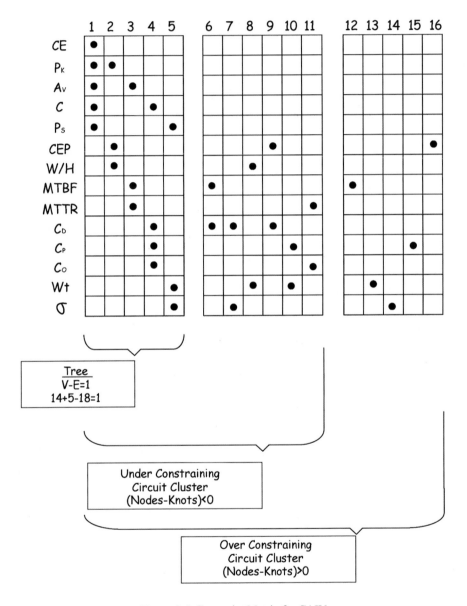

Figure 8-5. Constraint Matrix for CAIV

8.3 THE KINEMATICS OF FREE-FALL WEAPONS

Consider the traditional bombing problem: an aircraft attempts to attain a kinematic state which allows a free-falling bomb to impact on a target. A great diversity of avionic equipment exists to measure the aircraft kinematics, covering a great range of accuracy and cost. "What are the best variables to measure?" is a common question. Before this question can be properly answered, a more fundamental one must be addressed: "What sets of variables are necessary and sufficient for a bombing solution, i.e., exactly 'constrain' the trajectory?" That this latter question is nontrivial is shown in the following:

Define the following ten kinematic parameters of an aircraft with respect to its target (See Figure 8-6):

X	horizontal distance between aircraft and target
$\dot{X}$	horizontal velocity of aircraft
Z	vertical distance between aircraft and target
$\dot{Z}$	vertical velocity of aircraft
r	range from aircraft to target
$\dot{r}$	range rate, aircraft to target
S	angle from horizontal of range vector
$\dot{S}$	angle rate of range vector
V	aircraft velocity magnitude
d	aircraft velocity angle from horizontal.

For the sake of the example, only the kinematics within the bomb trajectory plane are considered. All the variables listed previously except X are measurable by at least one type of airborne instrument. We may now state the problem more specifically.

Which subsets of $(\dot{X}, Z, \dot{Z}, r, \dot{r}, S, \dot{S}, V, d)$ form complete descriptions of the aircraft in-plane kinematics? Call these describing sets.

From physical reasoning, the immeasurable set $(X, \dot{X}, Z, \dot{Z})$ completely defines the kinematics where (X, Z) define a two-dimensional positional vector, and $(\dot{X}, \dot{Z})$ define a two-dimensional velocity vector. Assuming for the moment that every describing set has four variables, then

$$C_4^9 = \frac{9!}{4!(9-4)!} = 126$$

four-element subsets of the measurable set must be examined to determine if it is a describing set. This can be an extremely tedious problem since only a few of these sets can be merely examined by inspection for easy classification as a describing set (Z, r, V, d), or not a describing set $(\dot{X}, \dot{Z}, \dot{V}, \dot{d})$

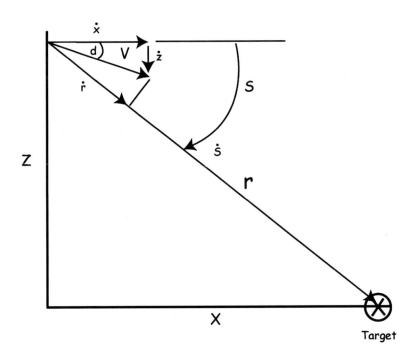

Figure 8-6. Weapon Delivery Variables (initial conditions).

Constraint theory provides both a point of view and a procedure to solve this type of problem whether it be ten-dimensional or 500-dimensional. First, a mathematical model which relates all the relevant variables is established. Second, meta models consisting of a bipartite graph and a constraint matrix are established to analyze the constraint properties of the original model. Third, these meta models are tested and adjusted, if necessary, for internal consistency. Fourth, the classification of any subset of measurables as a describing set or not a describing set can be made by systematically propagating constraint through the meta model. In this procedure, the subset to be analyzed acts as the input variables and any known describing set such as $(X, \dot{X}, Z, \dot{Z})$ acts as the output variables of a computational network.

Whether constraint theory is applied or not, the stated problem cannot be solved unless the couplings and interactions between the kinematic variables are clearly delineated as a set of relations. One such set may be the following:

1: $\quad r = (X^2 + Z^2)^{1/2}$
2: $\quad \dot{r} = (\dot{X}X + \dot{Z}Z)(X^2 + Z^2)^{-1/2}$
3: $\quad X = r \cos S$
4: $\quad Z = r \sin S$
5: $\quad S = \tan^{-1}(Z/X)$
6: $\quad \dot{S} = (\dot{Z}X - \dot{X}Z)(X^2 + Z^2)^{-1}$
7: $\quad V = (\dot{X}^2 + \dot{Z}^2)^{1/2}$
8: $\quad Z = -\dot{X} \tan d.$

Each relation is correct. More correct relations can be written, but the preceding list seems to be sufficient, whatever that means. Normally, the analyst must rely on his intuition or judgment regarding the point at which he should cease adding relations to his model. The bipartite graph and constraint matrix meta models are presented in Figure 8-7.

The fundamental criterion for math model consistency is that the total model relation set is not the null set. This simply means that the set of points in the total ten-dimensional space which satisfied every one of the eight relations must include least one point, otherwise some part of the model is intrinsically incompatible with another part. From the constraint point of view, model inconsistency is due to overlapping domains of intrinsic constraint.

Each relation in the example above is an n-1 dimensional regular surface where n is the dimension of the relation. Thus they are all regular relations, and the source of intrinsic constraint in this type of model is always a basic nodal square (BNS), a special subgraph of the total bipartite graph which is recognized by an equal number of vertices of each type. A logical first step in the consistency testing is to search the meta models for BNSs, thereby locating constraint sources.

Since for even this small model, a systematic search through all possible submodels involves examination of 256 cases, it is clear that more powerful means must be used to search for the BNS.

Every BNS is the union of simple nodal circuits. We know that every BNS must be a circuit cluster and can contain no treelike appendages. Examination of either Figures 8-7a or 8-7b reveals that the variables $\dot{r}$, $\dot{s}$, v, and d form trees attached to the main circuit cluster; therefore, the nodes 2, 7, 8, and 6 cannot be part of a BNS. Temporarily stripping these nodes ("trimming the trees") from the model, we get the circuit cluster depicted in Figure 8-8.

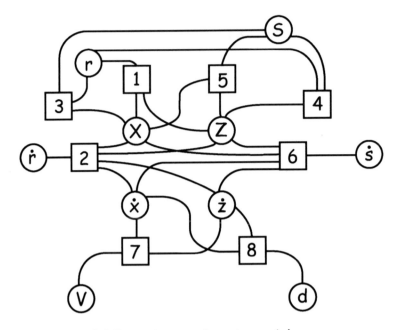

(a) Bipartite graph meta model.

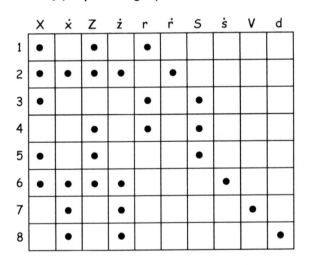

(b) Constraint matrix meta model.

Figure 8-7. Constraint Matrix Meta-model.

Since Figure 8-8 has the same number of each type of vertex (constraint potential equal zero), it is a nodal square. Since every node has a degree of three, there cannot exist smaller nodal squares of dimension 2x2 within

Figure 8-8. (The dimension of a nodal square must always be equal to or greater than the degree of any of its nodes.) Since the degree of every knot is also three, nodal squares of dimension 3x3 cannot exist within Figure 8-8 either. (In order for a nodal square of dimension n to contain a nodal square of dimension n-1, there must exist a knot of degree 1.) Therefore, Figure 8-8 is a nodal square which does not contain smaller nodal squares and is thus a basic nodal square. Moreover, it is the only BNS in the entire model.

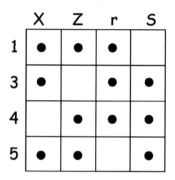

Figure 8-8. Circuit Cluster.

We can generally expect that a BNS subgraph of a model of regular relations will point constrain each of its variables. This does not violate the fundamental criterion for model consistency.

But let us examine this BNS in the context of the purpose of the model. We certainly did not expect the model to determine specific values of X, Z, r, and S. Yet this is just what four equations covering four unknowns threaten to do. Trying to resolve this absurdity, we focus our attention on this portion of the original model and discover that relations 3 and 4 are redundant to relations 1 and 5. 3 and 4 are therefore removed.

The new constraint matrix is shown in Figure 8-9. The total constraint potential (relations minus variables) is now equal to –4, which agrees with our physical reasoning that the total system should have 4 degrees of freedom. Moreover, the subgraph 1-5 dealing only with position, and the subgraph 7-8 dealing only with velocity, each have constraint potentials of –2, further agreeing with our physical notions of position and velocity vectors in a plane.

Certainly the set $(X, \dot{X}, Z, \dot{Z})$ is a describing set. Any other set (a, b, c, d) will also be a describing set if, by setting unique values for (a,b,c,d), we can compute unique values for $(X, \dot{X}, Z, \dot{Z})$. From the constraint matrix point of view, we apply extrinsic constraint at the four knots a, b, c, d, and allow the constraint to propagate throughout the matrix. If constraint propagates to $(X,$

$\dot{X}$, Z, $\dot{Z}$) and no other variable is over-constrained, then the trial set (a, b, c, d) is a describing set since it constrains the trajectory just as completely as (X, $\dot{X}$, Z, $\dot{Z}$).

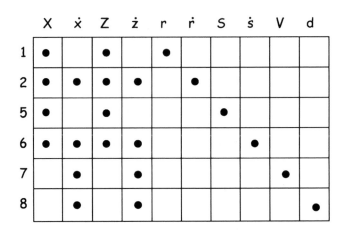

	X	$\dot{x}$	Z	$\dot{z}$	r	$\dot{r}$	S	$\dot{s}$	V	d
1	●		●		●					
2	●	●	●	●		●				
5	●		●		●			●		
6	●	●	●	●					●	
7		●		●	●				●	
8		●		●	●					●

Figure 8-9. Final Constraint Matrix.

Constraint propagation through a constraint matrix follows very simple rules which can be easily mechanized on a computer. When constraint is transmitted to a variable via one of its edges, it is transmitted through the variable via all its other edges. When constraint is transmitted to a relation via all but one of its edges, it is transmitted through the relation via its remaining edge. If this procedure does not completely fill the graph, then the residue is investigated for BNSs.

Using this procedure, the describing sets can be found rapidly. It will be more compact to list the measurable sets which are <u>not</u> describing sets. (See Table 8-3.)

Each of the remaining 99 sets of four measurables, even the unlikely $\dot{X}\dot{Z}\dot{r}\dot{S}$, $\dot{Z}\dot{r}\dot{S}V$, and $\dot{Z}\dot{r}\dot{S}d$, perfectly constrains the trajectory and is a describing set.

8.4　THE DEFLECTION OF AN EARTH-THREATENING ASTEROID EMPLOYING MASS DRIVERS

Since the early 1990's there has been an increasing awareness of the risk to earth by a strike from a large comet or asteroid. It is estimated that if such a near-earth object (NEO) of 1 km or larger struck the earth it would disrupt our global ecology and cause a billion human casualties. Much has been accomplished regarding the *detection* of these NEOs; the number of catalogued NEO's has grown from under 5% to over 50% of their estimated population from the early 90's to the early 00's.

Table 8-3. The set that are non-describing sets

These sets overconstrain velocity:

$\dot X Z \dot Z V$	$\ddot X \dot Z S V$	$\dot X r V d$
$\dot X Z \dot Z d$	$\ddot X \dot Z S d$	$\dot X S V d$
$\dot X Z V d$	$\dot X \dot Z \dot S V$	$\dot X \dot S V d$
$\dot X \dot Z r V$	$\dot X \dot Z S d$	$\dot Z r V d$
$\dot X \dot Z r d$	$\dot X \dot Z V d$	$\dot Z \dot r V d$
$\dot X \dot Z \dot r V$	$Z \dot Z V d$	$\dot Z S V d$
$\dot X \dot Z \dot r d$	$\dot X r v D$	$\dot Z \dot S V d$

These sets overconstrain position:

$\dot X Z r S$	$Z r \dot r S$	$Z r S V$
$Z \dot Z r S$	$Z r S \dot S$	$Z r S d$

However, much less attention has been paid to the *deflection* or *intercept* of these threats to human civilization. The methods which provide the highest energy density delivered to the threatening object are nuclear. But there is great resistance to testing and putting nuclear devices into space, the bomb may split the NEO in unpredictable ways, it may pollute the atmosphere, and the coupling of the energy to the NEO is critically dependent on the NEO's structure and composition.

The most robust and lowest technical risk approach is to employ traditional chemical propulsion to the NEO by attaching a conventional rocket to the object and transfer orbit-changing momentum. But the cost of this approach is enormous; not only must the chemical energy be delivered to the NEO, the reaction mass must be transported there also. Imagine a Saturn V rocket engine firing continuously for a year or more!

A concept which has received far less attention is to employ a mass driver – essentially a linear motor which converts electrical energy into kinetic energy. Once a mass driver is transported and installed on the NEO, the required energy can be obtained with a solar collector and the required reaction mass can be obtained from the NEO itself. The following analysis employs constraint theory to perform trade studies of mass driver designs for NEO deflection.

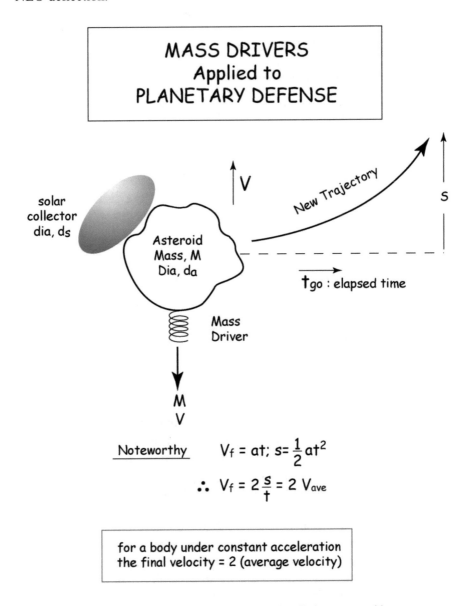

Figure 8-10. Schematic of Mass Driver installation on asteroid.

Refer to Figure 8-10 for a schematic of the mass driver installation on the NEO. The NEO, or asteroid, has mass M and diameter d_a. A solar collector of diameter d_s collects energy for the mass driver. The mass driver is a long coil which propels mass from the NEO in buckets with a velocity, v. The action of the mass driver ejecta causes a reaction by the asteroid, moving it in the opposite direction with velocity, V. In order to miss the earth within a time-to-go, t_{go}, the NEO must move a distance S – normally considered to be the earth's radius. (Generally, the optimum application of force for orbit changing is parallel to the asteroid's velocity vector. However this analysis stresses the case of final approaches with short time-to-go and small angles between the orbits of earth and asteroid.)

The kinematic relations can be captured by the bipartite graph in Figure 8-11. The variables are:

> S= distance moved perpendicular to the asteroid trajectory
> a= acceleration of asteroid
> t= time
> V_f= asteroid final velocity

and the relations are:

$$s=f_1(a,t)=(1/2)at^2 \quad V_f=f_2(a,t)=at$$

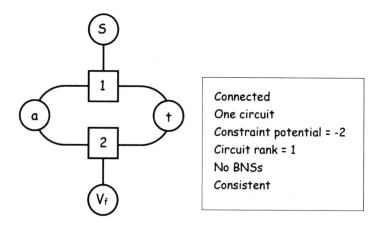

Connected
One circuit
Constraint potential = -2
Circuit rank = 1
No BNSs
Consistent

Figure 8-11. Bipartite Graph for initial kinematic model.

We can now determine whether computational requests can be made of this simple model:

The computational request, $v=f_3(a,s)$ is allowable and is: $v=\sqrt{2as}$ which is a quite familiar formula. The computational request, $v=f_4(s,t)$ is also allowable and it is: $V_f=2s/t$, but in the memory of the author, it was not familiar. He does not recall ever seeing this formula, and doesn't remember that, in a system undergoing constant acceleration with zero initial velocity, the final velocity is twice the average velocity. This was an unexpected bonus of this simple application of Constraint Theory. The value of the f_4 relation is that we can easily compute the total energy requirement to deflect the NEO from impact with the Earth, for any given time-to-go.

Let us now expand the model from kinematics to dynamics and write:

$$\text{Total energy imparted to the asteroid} = E_r = (1/2)\, MV_f^2 = 2M(V_{ave})^2$$

The energy available from the solar collector, $E_s = AFEt_{go}$, where:

A=the solar collector area
F=the solar flux density at that distance from the sun
E=the flux-to-electrical energy conversion efficiency
t_{go} = time-to-go

Expanding the bipartite graph to include these variables and relations yields Figure 8-12.

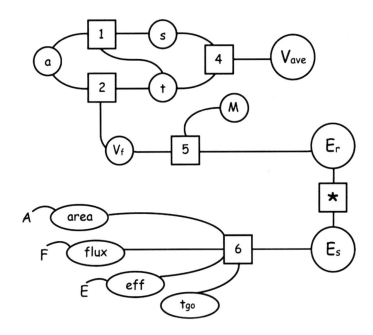

Figure 8-12. NEO Kinematics, NEO energy required and solar energy available.

It would be tempting to assume that all the energy collected by the solar array is available to provide the asteroid with the required energy, that is, the relation designated as "*" in Figure 8-12 is simply: $E_r = E_s$. Unfortunately, this is far from correct, since *both* energy and momentum relations must be satisfied:

Energy: $E_s=(1/2)MV_f^2+(1/2)mv^2$
Momentum: $MV_f=mv$

Define a new variable, $k=v/V=M/v$ – that is, the ratio of asteroid mass to ejecta mass, or equivalently, the ratio of ejecta velocity to asteroid velocity. Substituting k into the energy equation yields:

(Asteroid kinetic energy/ejecta kinetic energy)$=1/k$

In other words, only $1/(1+k)$ of the available solar energy couples to the asteroid, the remainder accelerates the ejecta.

Paradoxically, the situation gets worse as the ejecta velocity increases and the time-to-go increases. These results run counter to the "general wisdom" that high ejecta velocity is good – after all, much hard work has been expended in driving up the exhaust velocity of rocket engines and mass drivers – and that the more time we have for deflection, the better off we are. The general wisdom is still correct in their original domains and problem statements, but does not precisely apply to the issues of asteroid deflection.

All these results can be summarized on a single chart, Figure 8-13. Following the standard conventions employed to compare NEO deflection concepts, the two fundamental independent variables are time-to-go and asteroid diameter. The dependent variable is the diameter of the solar collector, d_s. At the center of the chart it can be seen that a solar collector of only 100 meter diameter is required to deflect a 1 kilometer asteroid if one year of time-to-go is available. The chart also presents an additional relation: the locus of $d_s=d_a$. On this locus, the solar collector is the same size as the asteroid and the time-to-go becomes the dependent variable.

In summary, although the number of variables of this model was not exceptionally high, the constraint theory methodology provided useful kinematic and dynamic insight, as well as managing the computational flow through a confusing set of equations.

Mass Drivers for Planetary Defense

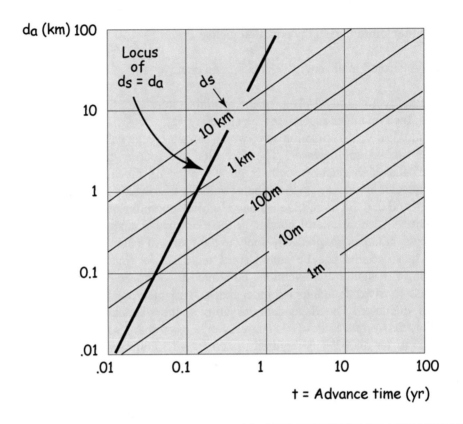

Asteroid Diameter (d_a) as a function of
Advance Time (t) and Solar Collector Diameter (d_s)

Compatible with Gehrels, "Hazards...p1119"

Figure 8-13. Results of the mass-driver analysis.

Chapter 9 MANAGER AND ANALYST MEET AGAIN

Gists and Schizophrenia

The analyst and manager met again to celebrate their newfound friendship and camaraderie.

"Would it be fair to ask if there is an overarching perspective to this book?" asked the manager, still a little overwhelmed.

"Not really fair," replied the analyst, "but I'll try:"

THE GIST OF IT ALL
Technology's ratchet is forcing systems into ever higher complexity.
Our greatest hope in managing complexity is math modeling.
But these models' dimensionality is incomprehensible to any human.
By performing computations on the model, we can optimize designs as
well as understand the systems' performance and phenomena.
But most of these computations are not allowable,
and it gets far worse with increasing dimension.
Computations are not allowable because they are not well-posed.
Allowable computations can only be made on a consistent model and
they must possess a proper constraint flow.
The kernel of constraint in a model is the basic nodal square (BNS)
and overlapping BNS's cause inconsistency and overconstraint.
The topological properties of the model's bipartite graph metamodel
provide practical clues for locating (and reconciling) BNS's.
Bipartite graphs are most useful for regular relations.

"Alternatively, another summary from a somewhat different viewpoint is presented in Figure 9-1," the analyst contributed.

© Springer International Publishing AG 2017
G.J. Friedman, P. Phan, *Constraint Theory*, IFSR International Series on Systems
Science and Engineering, DOI 10.1007/978-3-319-54792-3_9

"These are not very brief 'gists,'" complained the manager. "Haven't you heard the saying: 'That which is good and short is doubly good'?"

"Yes, I have," agreed the analyst, "but brevity depends on a shared experience – which we didn't have originally – and a deep understanding of the language we're using. Normally, the richer the language, the briefer the meaningful messages can become. If nothing else, this book has introduced you to new directions of mathematics, which is a special type of language."

"Speaking of language, it can reasonably be argued that constraint theory sits on a bridge between mathematics and qualitative language. All languages have built into their grammatical structure the primitive concepts of "none," "one" and "many." – I have no horse, I have one horse, I have many horses. The rhythm and pivotal concepts of constraint theory follow in analogous ways:

Regarding paths between parts of a bipartite graph:
- no paths: the parts are in disjoint connected components
- one path: the parts are connected by a tree-like structure
- many paths: the parts are within a simple circuit or a circuit cluster

Regarding circuit rank, c(G):
- $c(G) = 0$: no circuits exist in G
- $c(G) = 1$: one independent circuit must exist in G
- $c(G) > 1$: multiple independent circuits must exist in G

Regarding constraint potential, p(G):
- $p(G) \geq 0$: BNS must exist in G.

"But it's regrettable, isn't it, that the bipartite graph couldn't be as useful for discrete and interval relations as it is for regular relations?"

"I do wish it would be more broadly applicable," agreed the analyst, but can you name me *any* branch of mathematics that is equally useful over all applications? Math tends to be very specialized into specific application domains. Consider the vast body of math that analyzes linear systems, despite the fact that most real-world systems are nonlinear. On the bright side, regular relations presently represent the most common type of relation used in math modeling, so if it's limited to a domain, at least it's the most significant domain."

"As I promised in Chapter One," continued the analyst, "Constraint Theory really occupies just a simple portion of all mathematics. Consider John Barrow's 'Structure of Modern Mathematics' [16] shown in Figure 9-1, where Constraint Theory is represented only in the simplest lower right-hand corner. We made substantial use of bipartite graphs which are with graph theory and closely related to hypergraphs since the constraint matrix of a bipartite graph is essentially identical to the incidence matrix of a hypergraph [17]."

"There is another bright side," the analyst added. "This book is only the first venture into a new world of applied mathematics, and I'm hopeful that far more research will be accomplished in the future."

"I hope so too. We've been through a lot together," commented the manager, "more than I had expected at the beginning."

"Yes," agreed the analyst, "our interactions seemed to take on a life of their own, far exceeding the expectations from the original seeds of our discussions. In a way, that's a testimony to the richness and importance of the area."

"Speaking of richness, you seem to have had a great diversity of prior experience to provide the basis for such an invention as constraint theory," complemented the manager.

"Well yes, I suppose I have," admitted the analyst. "I had the good fortune to have worked in the fields of systems, mechanical, electrical and civil engineering, applied mathematics, computer science, control systems, surveying, soil mechanics, structural design, rockets, guidance and control, instrumentation, steam power plants, environmental systems, pressure sensors, accelerometers, altimeters, air data computers, missile fueling systems, stellar inertial navigation, electro-optical sensors, acoustic terminal homing sensors, optical gyroscopes, electronic countermeasures, airborne radar, artificial intelligence, and low observables technology, to name a few."

"Very impressive," beamed the manager. "I too have had the good fortune to be involved with a great variety of programs, including the V-2 ballistic missile, the Redstone, Jupiter and Thor missiles, the Baldwin Hills reservoir, the San Fernando Valley Steam Electric Plant, the Skybolt air-launched ballistic missile, the F-5 fighter aircraft, the Advanced medium range air-to-air missile, the brilliant antitank submunition and the B-2 stealth bomber. In fact, I believe I'm the only person to have worked on both the V-2 and the B-2. Additionally, I have contributed to workshops for all branches of the US Department of Defense, NASA, DOE, NSF and NATO."

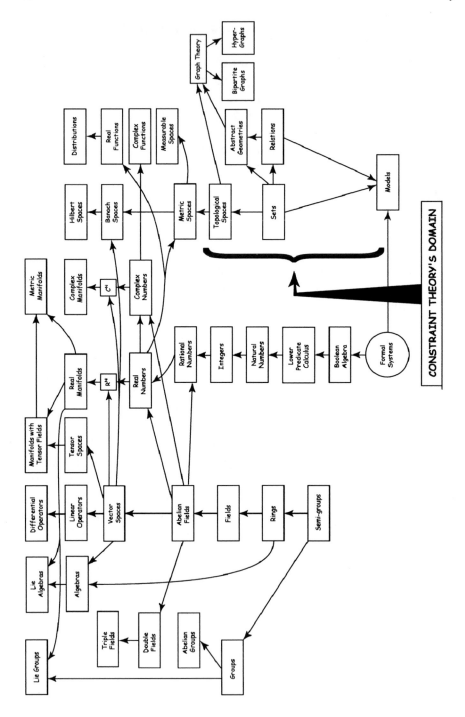

Figure 9-1. The Structure of Modern Mathematics
(From: John D. Barrow, The Book of Nothing, Pantheon Books, 2000, p152)

"Do you think it a coincidence that the list of technologies are generally relevant to the list of programs?" asked the analyst.

"Not if you realize that we're really the same guy," both observed. "And the dialogues we used were merely an expository mechanism to describe a new field and illuminate unfamiliar concepts. We were attempting – without daring to be compared to them – to use dialogue in the manner of Plato, Galileo and even Mark Twain in his philosophical work, 'What is Man?'

> *Roses are red,*
> *violets are blue.*
> *I am schizophrenic,*
> *and so am I.*

Appendix A COMPUTATIONAL REQUEST DISAPPOINTMENTS; RESULTS OF THE USC ALLOWABILITY PROJECT

In order to construct a useful and trustworthy mathematical model, one must gather authoritative data and relations, ask the advice of experts in the domains of each submodel, assure that all uses of each variable agrees semantically, and laboriously imbed the structure into a computer. After all this painstaking work, it would be hoped that one could derive new understandings across the many domains of the submodels and exercise wide liberty in making computational requests of the math model.

It has been said of Carl Friedrich Gauss [19] that he "achieves greatness in his work not through deep, abstract mathematical thinking, but rather through an incredible vision of how the various quantities in the problem are related, a vision that guides him through extraordinary computations that others would likely abandon as futile."

As was thoroughly discussed in the main text, a crucial impediment to the allowability of computational requests is that the model is not consistent due to intrinsic overconstraint. Clearly, no computational requests are allowable on an inconsistent model. If this inconsistency is removed by eliminating an appropriate number of submodels there still may exist intrinsic basic nodal squares (BNSs) which apply constraint to all their relevant variables.

Although the model has been rendered consistent, none of the variables relevant to the BNSs can participate as an independent variable, dependent

G.J. Friedman, P. Phan, *Constraint Theory*, IFSR International Series on Systems Science and Engineering, DOI 10.1007/978-3-319-54792-3

variable, or variable held constant. This significantly reduces the number of allowable computational requests.

Unfortunately, even if <u>all</u> the BNSs were to be removed – making the model completely free of intrinsic overconstraint and point constraint, the likelihood of any given computational request being allowable is still quite low. Examples may clarify this disappointing fact.

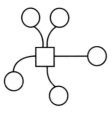

Consider the trivial case for N=1; the model has only one node and K knots. There are 2^K possible computational requests, but only one allowable computation (identical to the original contributing submodel.) Thus, the probability of allowability, p(a) equals $1/2^K = 2^{-K}$. If K=4, then p(a) = 1/16.

Next, consider a model with N=2 having a tree structure and with K_1 variables relevant to N_1 and K_2 variables relevant to N_2. Since one of the variables is shared, the total number of variables is K_1+K_2-1 and the total number of possible computational requests = $2^{K_1+K_2-1}$. The total allowable computational requests are the two contributing submodels, plus K_1-1, plus K_2-1, which merely sums to K_1+K_2. Thus, for this case,

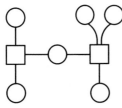

$$p(a) = \frac{(K_1 + K_2)}{2^{K_1+K_2-1}}$$

If $K_1=K_2=4$, then p(a) = 1/16 again.

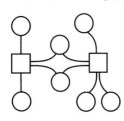

If the N=2 model has a circuit structure with a single circuit, then two of the knots are shared and the total knots in the model = K_1+K_2-2. The total number of allowable computations here is only 4, so in this case, $p(a) = \dfrac{4}{2^{K_1+K_2-2}}$, and if the K's = 4, p(a) = 1/16 (again!!)

This remarkable coincidence of p(a) = 1/16 does not extend to more complex models. Continuing this mode of analysis becomes exceedingly complex with higher N and K; the number of possible computational requests for a simple model of only N=10 already exceeds 1000. Thus let us skip ahead a bit to the region of K=6,8 and N=4,6 which was employed in Chapter 1 to demonstrate how even these low dimensions can exhibit interesting behavior. We will employ the results of the *USC Computational Allowability project* [1] wherein three engineering graduate students employed the techniques of Chapters 4 and 5 and exhaustively performed allowability analyses on the models shown below.

In the analyses to follow, a key "cognitive limitation" was placed on the total number of computational requests to be considered: only 2, 3 and 4 dimensional requests were analyzed. (For example, $x=f(y,z)$ is defined as a three dimensional request.) This limitation was employed for two reasons:

(a) most importantly, dimensions of 5 and higher require complex "carpet plots" to understand and are very difficult for most people to perceive, and

(b) with the constraint potential of the examples equaling -2, it is highly unlikely that computational requests of dimension higher than 4 would be allowable anyway.

Figure A-1. Bipartite Graph Topological Structures Examined.

Thus, instead of examining all 2^K conceivable computational requests we only examine $\dfrac{K!}{2!(K-2)!} + \dfrac{K!}{3!(K-3)!} + \dfrac{K!}{4!(K-4)!}$ of them. For K=6, only 50 out of the total possible of 64 were examined, for K=8, only 154 out of the total possible of 256 were examined.

Figure A-1 summarizes the ten bipartite graph topological structures which the team examined. Table A-1 summarizes the results of the 1228 different computational requests which were made on these ten topologies.

Table A-1. Summary of Computational Request Allowability Analysis

Struct-ure	K	N	E	C	B	# Examined	# Allowable	% Allowable
A	6	4	12	3	0	50	13	26%
B	6	4	11	2	0	50	12	24
C	6	4	10	1	0	50	9	18
D	8	6	18	5	1	154	6	4
E	8	6	18	5	0	154	23	15
F	8	6	17	4	1	154	6	4
G	8	6	16	3	1	154	6	4
H	8	6	15	2	1	154	6	4
J	8	6	14	1	1	154	3	2
K	8	6	13	0	0	154	6	4
					TOTALS	1228	90	7.3%

LEGEND:

Structure:	Bipartite graph topological structure shown in Figure A-1
K:	Number of Knots, or variables
N:	Number of Nodes, or relations
E:	Number of Edges, or relevancies
C:	Number of independent circuits, or circuit rank of the graph
B:	Number of Basic Nodal Squares
# Examined:	Number of computational requests which were analyzed
# Allowable:	Number of requests which were computationally allowable
%:	Percent of the requests which were allowable

The conclusion is rather dramatic: the overall probability of allowability was only 7.3%. If the analysis had been extended to all 2^K computational requests, the percent allowability would have been much less.

Not surprisingly, those topologies which were devoid of BNSs had the highest allowabilities, but even then it was only about 20%. Therefore it should be abundantly clear that an allowability analysis must be performed on every computational request before the programmers jam it into the machines.

Worse yet, it is the author's conjecture that – as the model dimensionality increases to 100, 1000 and beyond – the absolute number of allowable computational requests will increase, <u>but</u> the percentage of allowability based on the total number of possible requests will <u>decrease</u> as illustrated in Figures A-2 and A-3.

Some things are easier to prove
with the
CONSTRAINT MATRIX:

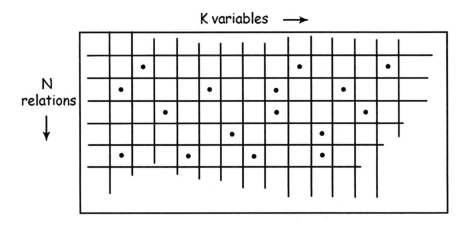

Each of the NK elements either has a " • " or not; thus:

> The number of possible topologies
> of a model with K variables and
> N relations = 2^{NK}

For a model with: $\left\{ \begin{array}{l} \text{K variables} \\ \text{N relations} \\ \text{(N } \sim \text{ K-2)} \end{array} \right.$

Conjecture:
The number of allowable computational
requests, A, is between K and K^{10}
[sounds very conservative, but am
I making an "Archimedes error"?]

Proof:
The number of total possible computational
requests, T, equals the number of
topologies times the number of requests
per topology:

$$T = \underbrace{2^{KN}}_{\text{topologies}} \underbrace{2^{K}}_{\substack{\text{requests /} \\ \text{topology}}} \approx 2^{K(K-1)}$$

Consequences $\longrightarrow$

$$K < A < K^{10}$$
$$T = 2^{K(K-1)}$$

K	A	K(K-1)	$T=2^{K(K-1)}$	A/T
10	10^{10}	90	$2^{90} \sim 10^{27}$	10^{-17}
100	10^{20}	9900	$2^{9900} \sim 10^{3000}$	10^{-2980}
1000	10^{30}	999000	$2^{999000} \sim 10^{300000}$	$10^{-299,970}$

Figure A-2. Typical values of A&T for K = 10, 100, 1000

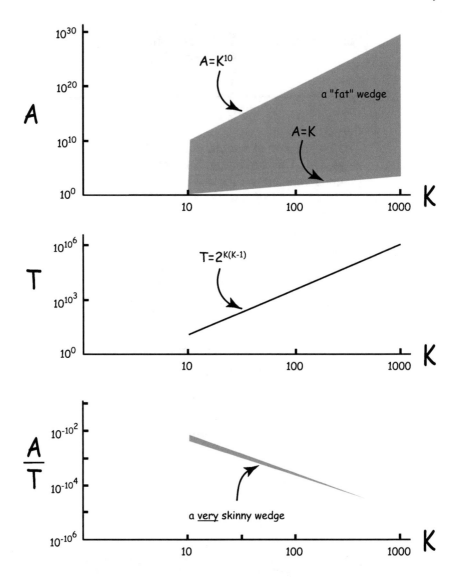

Figure A-3. A, T, & A/T vs. K

Appendix B **GRAPH THEORY OVERVIEW**

Why was the Bipartite Graph Chosen?

A *general graph,* or *graph,* consists of a set of vertices, some pairs of which are connected by a set of *arcs,* or *edges.*

A *bipartite graph (BPG)* is a graph such that its vertices can be decomposed into two disjoint sets, X and Y, and its edges only connect vertices in set X to vertices in set Y.

Any property of a general graph is also a property of a bipartite graph, but not vice versa (Table B-1). For example, both the general and bipartite graphs have connected components, trees and circuit structures. However only the bipartite graph has the property of constraint potential and the structure of basic nodal squares.

Graphs have been used for many decades as mathematical metamodels. Mason, et al [20], applied *signal flow graphs (SFG)* to the analysis of complex systems where the functions were linearly separable. Friedman [21] applied *inverse signal flow graphs ($\overline{SFG}$)* to a broader class of systems. Figure B-1 provides a comparison between how the signal flow graph, the inverse signal flow graph and the bipartite graph assign portions of their structure to the mathematical concepts of variable, relation and relevancy.

	SFG	$\overline{SFG}$	BPG
Variable	Vertex	Edge	Knot
Relation	Edge	Vertex	Node
Relevancy	Edge	Edge	Edge

Figure B-1. The definitions of Three Types of Graph Theory Metamodels.

© Springer International Publishing AG 2017
G.J. Friedman, P. Phan, *Constraint Theory*, IFSR International Series on Systems Science and Engineering, DOI 10.1007/978-3-319-54792-3

For the purposes of Constraint Theory, the bipartite graph metamodel was chosen over the other alternatives because it is more general topologically and functionally, as well as the fact that it can represent *both* the model and the computations performed on it.

Figure B-2 demonstrates the greater generality of the BPG by showing that any SFG can be represented by a BPG, but most BPGs cannot be represented by a SFG or an $\overline{\text{SFG}}$. Figure B-3 demonstrates the greater generality of the BPG by showing that the SFG is limited only to linearly separable functions where the BPG can represent *any* function.

Table B-1. Graph Theory Overview.

<u>General Graph</u>: A set of vertices, some pairs of which are connected by a set of arcs (or edges)

<u>Bipartite Graph</u>: The vertices have two species and the arcs connect only two different species

Any property of a general graph is also a property of a bipartite graph, but not vice versa

Graphs have been used for many decades to describe computations in complex networks; i.e.; Signal Flow Graphs

The Bipartite Graph was chosen over the Signal Flow Graph because it is more general topologically and functionally and because it represents *both* model and computation

The BPG is more general TOPOLOGICALLY :

Any SFG can be represented by a BPG :

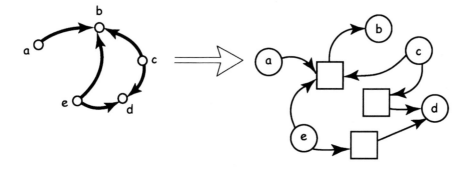

But most BPG's can't be represented by SFG's :

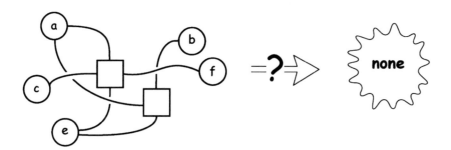

Figure B-2. BPG Characteristics (1).

The BPG is more general FUNCTIONALLY :

$a = k_1 c + k_2 d$ $b = k_2 c + k_4 d$	$a = k_5 cd$ $b = k_6 cd$	$a = d \displaystyle\int_0^c \sin(t)\, dt$ $b = \cos(cd)$
BPG : 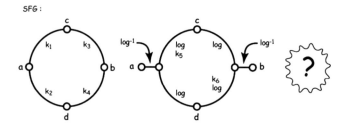	← ditto	← ditto

SFG :

Figure B-3. BPG Characteristics (2).

Appendix C THE LOGIC OF "IF" AND "IF AND ONLY IF"

Theorem 1 states that inconsistency implies computational non-allowability. More compactly, we can state: "nonallowability *if* inconsistent." If the reverse, "inconsistent *if* nonallowable" were true then we could state, "nonallowability *if and only if* inconsistent." (For even increased compactness, "*if and only if*" is often contracted to "*iff*".)

Actually, Theorem 1 is an "if" statement, not an "iff" statement. If a model is inconsistent, then no computation on it is permitted. However, the reverse is not always true because there are many other possible computability problems even with consistent models. Actually, the vast majority of the theorems presented in Chapters 3 and 4 are "if" theorems, not "iff" ones. Another example of a one-way "if" theorem is Theorem 25 which states that a BNS must always lie within a circuit cluster. Many students then wrongfully assumed that – whenever they identified a circuit cluster—they have found a BNS.

On the other hand, Theorems 13 and 14, dealing with the topological properties of trees and circuits, are indeed two-way "iff" theorems.

In order to appreciate the logic of "if" and "iff" statements more deeply, let us examine Figure C-1 which presents a Venn Diagram that displays the events of consistency, allowability and their interactions.

© Springer International Publishing AG 2017
G.J. Friedman, P. Phan, *Constraint Theory*, IFSR International Series on Systems Science and Engineering, DOI 10.1007/978-3-319-54792-3

Define these events:

    A: set of circumstances where the model is consistent
not A: set of circumstances where the model is not consistent
    B: set of circumstances where the computation is allowable
not B: set of circumstances where the computation is not allowable

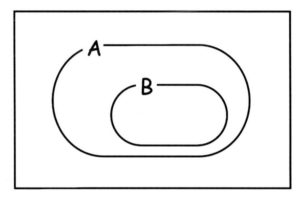

Therefore: B ⟶ A, and not A ⟶ not B
However, we <u>cannot</u> conclude that A ⟶ B, or not B ⟶ not A

Figure C-1. Logical Analysis of Theorem 1.

Appendix D ALGEBRAIC STRUCTURES

DEFINITIONS AND PROPERTIES OF GENERAL VECTOR SPACES

(Gross & Yellen, 2006, pp. 687 & 690; Anton, 1977, p. 127)

Definition D-1: A ***binary operation*** * on a non-empty set A is a function f: A × A → A, given by f [(a, b)] = a * b. The set A together with a binary operation * is denoted (A, *).

Definition D-2: The binary operation * on set A is said to be ***associative*** if for a, b, c ∈ A, (a * b) * c = a * (b * c), and ***commutative*** if for a, b ∈ A, a * b = b * a.

Definition D-3: An element e ∈ A is an ***identity element*** in (A, *) if for all a ∈ A, a * e = e * a = a. And for all a ∈ A, an element a' ∈ A is an ***inverse*** of a in (A, *) if a * a' = a' * a = e.

Definition D-4: A ***group*** G = (G, *) is a non-empty set G and a binary operation * that satisfy the following conditions:
  ➢ The operation * is associative.
  ➢ G has an identify element.
  ➢ Each g ∈ G has a unique inverse in (G, *), denoted g^{-1}.

Definition D-5: An ***abelian*** group is a group whose operation is commutative. In an abelian group, the binary operation is commonly denoted "+" and called *sum*.

Definition D-6: A ***field*** F = (F, +, •) is a set F together with two operations, + and • (generically called *addition* and *multiplication*), that meet the following conditions:
  ➢ (F, +) is an abelian group.
  ➢ (F − {0}, •) is an abelian group, where 0 is the *additive identity*.

© Springer International Publishing AG 2017
G.J. Friedman, P. Phan, *Constraint Theory*, IFSR International Series on Systems
Science and Engineering, DOI 10.1007/978-3-319-54792-3

➤ $a \bullet (b + c) = (a \bullet b + a \bullet c)$ and $(a + b) \bullet c = (a \bullet c) + (b \bullet c)$.

Definition D-7: The **_finite field GF(2)_** consists of the set $\mathbf{Z}_2 = \{ 0, 1 \}$ together with the mod-2 operations $+_2$ and $\bullet$. Thus:

➤ $0 +_2 0 = 1 +_2 1 = 0$ (this operation is identical to the logical operation of *exclusive OR*)
➤ $0 +_2 1 = 1 +_2 0 = 1$
➤ $0 \bullet 0 = 1 \bullet 0 = 0 \bullet 1 = 0$
➤ $1 \bullet 1 = 1$

Definition D-8: A **_vector space_** over a field (of *scalars*) F is a set V (of *vectors*) together with an operation $+$ on V and a mapping, called *scalar multiplication* from the Cartesian product $F \times V$ to V ($(a, \mathbf{v}) \rightarrow a\mathbf{v}$), such that the following conditions are satisfied for all scalars a, b $\in F$ and all vectors $\mathbf{v}, \mathbf{w} \in V$:

➤ (V, $+$) is an abelian group, where the notation "$+$" is being used to denote both addition of scalars in field F and addition of vectors in set V.
➤ $(a \bullet b) \mathbf{v} = a (b\mathbf{v})$
➤ $(a + b) \mathbf{v} = a\mathbf{v} + b\mathbf{v}$
➤ $a (\mathbf{v} + \mathbf{w}) \mathbf{v} = a\mathbf{v} + a\mathbf{w}$
➤ $e\mathbf{v} = \mathbf{v}$, where e is the *multiplicative identity* of field F.

BINARY SET OPERATIONS (Gross & Yellen, 2006, pp. 197-198)

Definition D-9: Let $s_1, s_2, \ldots, s_n$ be any sequence of objects, and let A be a subset of $S = \{ s_1, s_2, \ldots, s_n \}$. The **_characteristic vector_** of subset A, denoted charvec (A), is the n-tuple whose j^{th} component is 1 if $s_j \in A$, and 0 otherwise.

In the model graph G of Figure D-1 below, $E_G = \{ e_1, e_2, e_3, e_4, e_5, e_6, e_7 \}$. Cycle A has as its edge set $E_A = \{ e_1, e_2, e_3, e_4 \}$, and cycle B has as its edge set $E_B = \{ e_3, e_5, e_6, e_7 \}$, both of which are subsets of E_G. By **_Definition D-9_** above, charvec (E_A) = (1, 1, 1, 1, 0, 0, 0), and charvec (E_B) = (0, 0, 1, 0, 1, 1, 1).

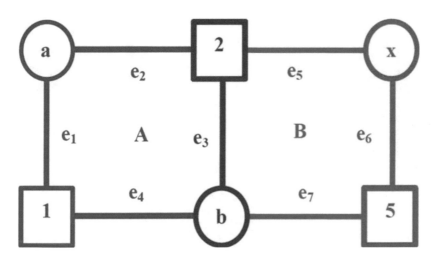

Cycle	Characteristic Vector of Edge Set	Edges						
		e_1	e_2	e_3	e_4	e_5	e_6	e_7
A	charvec (E_A)	1	1	1	1	0	0	0
B	charvec (E_B)	0	0	1	0	1	1	1

Figure D-1. Cycles and characteristic vectors of their edge sets

Definition D-10: For graphs A and B, the **union** of their edge sets E_A and E_B is the set of all edges which are either in E_A or in E_B, or both. Symbolically, $e \in E_A \cup E_B$ if $e \in E_A$ or $e \in E_B$.

In Figure D-2 below, edge set $E_A = \{ e_1, e_2, e_3, e_4 \}$ and edge set $E_B = \{ e_3, e_5, e_6, e_7 \}$. Thus, their union set $E_A \cup E_B = \{ e_1, e_2, e_3, e_4, e_5, e_6, e_7 \}$. Note that charvec($E_A \cup E_B$) is formed by combining the respective components of charvec(E_A) and charvec(E_B) with the bitwise **inclusive OR** operator. In C/C++, this operator has the syntactical symbol of "$|$", e.g. $(0 | 0) = 0$; and $(0 | 1) = (1 | 0) = (1 | 1) = 1$. In this case, charvec $(E_A \cup E_B) = (1, 1, 1, 1, 1, 1, 1)$.

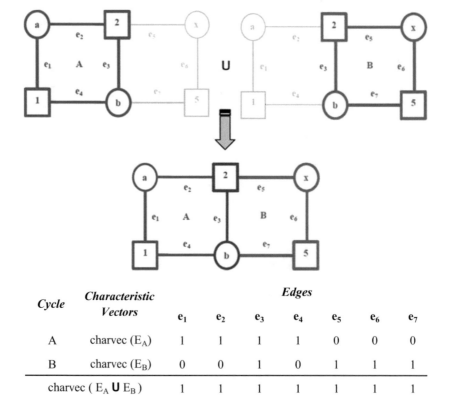

Cycle	Characteristic Vectors	Edges						
		e_1	e_2	e_3	e_4	e_5	e_6	e_7
A	charvec (E_A)	1	1	1	1	0	0	0
B	charvec (E_B)	0	0	1	0	1	1	1
charvec ($E_A \cup E_B$)		1	1	1	1	1	1	1

Figure D-2. Union of edge sets and its characteristic vector

Definition D-11: For graphs A and B, the **intersection** of their edge sets E_A and E_B is the set of all edges which are in both E_A and E_B. Symbolically: $e \in E_A \cap E_B$ if: $e \in E_A$ and $e \in E_B$.

In Figure D-3 below, edge set $E_A = \{\ e_1, e_2, e_3, e_4\ \}$ and edge set $E_B = \{\ e_3, e_5, e_6, e_7\ \}$. Thus, their intersection set $E_A \cap E_B = \{\ e_3\ \}$. Note that charvec ($E_A \cap E_B$) is formed by combining the respective components of charvec (E_A) and charvec (E_B) with the logical operator **AND** where 1 means true, and 0 means false. Symbolically, (0 *AND* 0) = (0 *AND* 1) = (1 *AND* 0) = 0; (1 *AND* 1) = 1. In this case, charvec ($E_A \cap E_B$) = (0, 0, 1, 0, 0, 0, 0).

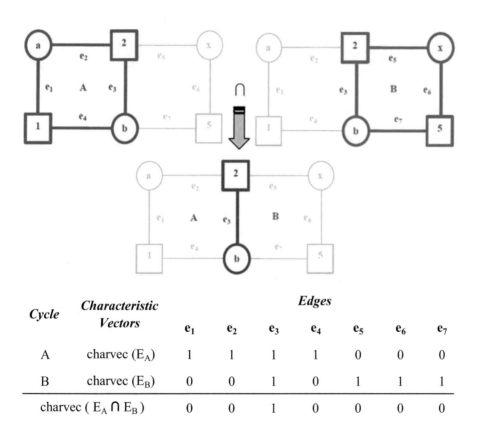

Cycle	Characteristic Vectors	Edges						
		e_1	e_2	e_3	e_4	e_5	e_6	e_7
A	charvec (E_A)	1	1	1	1	0	0	0
B	charvec (E_B)	0	0	1	0	1	1	1
charvec ($E_A \cap E_B$)		0	0	1	0	0	0	0

Figure D-3. Intersection of edge sets and its characteristic vector

Definition D-12: For graphs A and B, the **_difference_** $E_A - E_B$ is the set of all edges which are in E_A but **not** in E_B. Note that the difference is not necessarily commutative, i.e. ($E_A - E_B$) $\neq$ ($E_B - E_A$).

As demonstrated in Figure D-4 below, edge set $E_A = \{ e_1, e_2, e_3, e_4 \}$ and edge set $E_B = \{ e_3, e_5, e_6, e_7 \}$, therefore the difference set $E_A - E_B = \{ e_1, e_2, e_4 \}$. However, the difference set $E_B - E_A = \{ e_5, e_6, e_7 \}$.

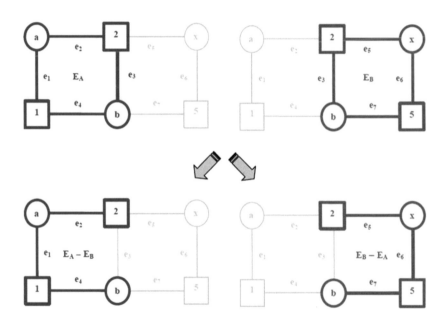

Figure D-4. Difference set ($E_A - E_B$) $\neq$ difference set ($E_B - E_A$)

Definition D-13: For a graph G, let $W_E(G)$ denote the set of all subsets of E_G, i.e. the power set of all edges in G. The **_ring sum_** of two elements E_1, E_2 $\subset W_E(G)$ is defined as:

$$E_1 \oplus E_2 = (E_1 - E_2) \cup (E_2 - E_1)$$

Figure D-5 below illustrates graphically an application of ring sum in which two cycles A and B are combined to form a new cycle. Note that charvec ($E_A \oplus E_B$) is formed by combining the respective components of charvec (E_A) and charvec (E_B) with the bitwise **_exclusive OR_** operator. In C/C++, this operator has the syntactical symbol of " ^ ", e.g. ($0 \wedge 0$) = ($1 \wedge 1$) = 0; and ($0 \wedge 1$) = ($1 \wedge 0$) = 1.

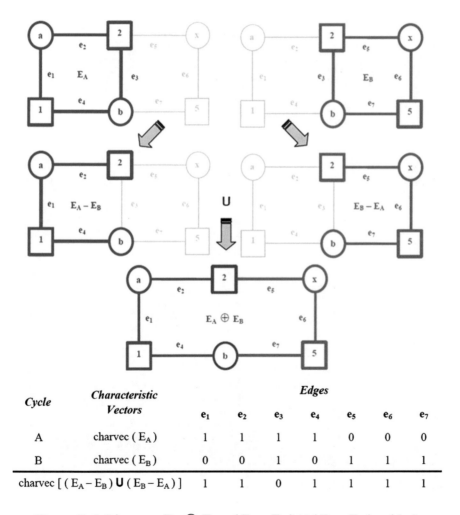

Cycle	Characteristic Vectors	Edges						
		e_1	e_2	e_3	e_4	e_5	e_6	e_7
A	charvec (E_A)	1	1	1	1	0	0	0
B	charvec (E_B)	0	0	1	0	1	1	1
charvec [($E_A - E_B$) $\cup$ ($E_B - E_A$)]		1	1	0	1	1	1	1

Figure D-5. Ring sum $E_A \oplus E_B = (E_A - E_B) \cup (E_B - E_A) = (1, 1, 0, 1, 1, 1, 1)$

References

1. Leila Habibibabadi, Kathya Zamora-Diaz, Elliott Morgan, *Computational Allowability on Bipartite Graphs*, projects for the graduate course, Systems Engineering, ISE 541, University of Southern California, 1998, 1999.

2. George Friedman, *Constraint Theory Applied to Mathematical Model Consistency and Computational Allowability*, PhD Dissertation, University of California at Los Angeles, 1967. University Microfilms, Inc, Ann Arbor, MI

3. George Friedman, *Constraint Theory, an Overview*, International Journal of Systems Science, 1976, Vol 7, No 10, pp 1113-1151.

4. Edgar Palmer, Graphical Evolution, John Wiley and Sons, New York, 1985

5. Roald Hoffmann and Shira Schmidt, Old Wine, New Flasks, W. H. Freeman and Company, New York, 1997

6. Norbert Wiener, A Simplification of the Logic of Relations, *Proceedings of the Cambridge Philosophical Society*, Vol 17, pp 387-390, 1914

7. Claude E. Shannon, The Theory and Design of Linear Differential Equation Machines, *National Defense Research Committee Report*, Princeton University, Jan 1942

8. George Gamow, One, Two, Three...Infinity, Viking Press, New York, 1947

9. Bourbaki, Elements de Mathematique; Theorie des Ensembles, ASEI 1141; Hermann & Cie, Paris, 3me. ed., 1958

© Springer International Publishing AG 2017
G.J. Friedman, P. Phan, *Constraint Theory*, IFSR International Series on Systems Science and Engineering, DOI 10.1007/978-3-319-54792-3

10. W. Ross Ashby, The Set Theory of Mechanism and Homeostasis, *General Systems, Yearbook of the Society for General Systems Research* Volume IX, Bedford, MA 1964

11. Edwin A. Abbott, Flatland, A Romance of Many Dimensions, Harper Collins, 1994; (Originally published circa 1880)

12. Dionys Burger, Sphereland, A Fantasy about Curved Spaces and an Expanding Universe, Harper Collins, New York, NY, 1994

13. K. Devlin, The Math Gene, Basic Books, New York, NY, 2000

14. P. J. Hall, *Representation Theorem in Set Theory*, J. Lond. Math. Soc, 10, 26. 1934

15. Antonin Svoboda, "An Algorithm for Solving Boolean Equations," *IEEE Trans. Electron. Comput*, Oct 1963, pp 557-559.

16. John Barrow, The Book of Nothing, Pantheon Books, 2000

17. Claude Berge, Hypergraphs, North Holland, 1989

18. T. Gehrels, *Hazards due to Comets and Asteroids*, University of Arizona Press, 1995

19. Ivars Peterson, Math Trek, *Science News Online*, April 17, 1999

20. S. J. Mason, Feedback Theory -- Some Properties of Signal Flow Graphs *Proceedings of the IRE*, Sept 1953, vol 41, pp 1144-56

21. G. J. Friedman, Constraint Algebra -- a Supervisory Programming Technique and a Cognitive Process, *IEEE Transactions on Military Electronics*, April 1963, pp 163- 167

22. Phan, Phan (2011). *Expanding constraint theory to determine well-posedness of multi-dimensional math models* (Doctoral dissertation). University of Southern California.

23. Tarjan, Robert Endre (1972). Depth-first search and linear graph algorithms. *SIAM Journal on Computing,* Vol. 2, pp. 146-160.

24. Cormen, T. H., C. E. Leiserson, R. L. Rivest and C. Stein (2001). *Introduction to Algorithms*. Cambridge, MA: The MIT Press.

25. Dechter, Rina (2003). *Constraint Processing*. San Francisco, CA: Morgan Kaufmann Publishers.

26. Gross, Jonathan L. and Jay Yellen (2006). *Graph Theory and Its Applications*. Boca Raton, FL: Chapman & Hall/CRC.

27. Tarjan, Robert Endre (1974, March). A note on finding the bridges of a graph. *Information Processing Letters,* Vol. 2, No. 6, pp. 160-161.

28. Shirey, R. W. (1969). *Implementation and analysis of efficient graph planarity testing algorithms* (Doctoral dissertation). University of Wisconsin, Madison, WI.

29. Anton, Howard (1977). Elementary Linear Algebra. New York, NY: John Wiley & Sons, Inc.

Index

© Springer International Publishing AG 2017
G.J. Friedman, P. Phan, *Constraint Theory*, IFSR International Series on Systems
Science and Engineering, DOI 10.1007/978-3-319-54792-3

Printed in the United States
By Bookmasters